Table of Contents

1 Chapter 1: Acoustic Origins

Sound is, at its core, a vibration. It's a mechanical wave, a disturbance that propagates through a medium, be it air, water, or even solid materials. This wave is created by a source – a struck drum, a vibrating string, a human voice – that sets the surrounding particles into motion. These particles then collide with their neighbors, transferring the energy and creating a chain reaction that travels outward from the source. Unlike electromagnetic waves, which can travel through a vacuum, sound requires a medium to propagate. Consider a scenario: A tree falls in a forest with no one around to hear it. Does it make a sound? The answer, from a purely physics perspective, is yes. The falling tree creates vibrations in the surrounding air, generating a sound wave. However, without a receiver (an ear, a microphone) to detect and interpret those waves, there is no perceived sound, no auditory experience. The speed at which sound travels depends on the properties of the medium. In air at room temperature, sound travels at approximately 343 meters per second (767 mph). However, in water, sound travels much faster, at around 1480 meters per second, and even faster in solids like steel. This difference in speed is due to the density and elasticity of the medium. Denser and more elastic materials allow sound waves to propagate more quickly. The way the wave spreads out is also worth noting, it's far from a uniform expansion and it depends on the environment it propagates through.

The anatomy of a sound wave is crucial to grasp. We often visualize sound waves as sinusoidal curves, representing the cyclical variations in pressure. The peaks of these curves represent areas of compression, where the particles are tightly packed together, while the troughs represent areas of rarefaction, where the particles are spread further apart. The distance between two consecutive peaks (or troughs) is called the wavelength, which is inversely proportional to the frequency of the sound. A shorter wavelength corresponds to a higher frequency, and vice versa. Understanding the spatial properties of sound is also very important. For example, a sound can be perceived as louder if the receiver of the sound is closer to the source of the sound, because the wave is more powerful there. It's possible to create devices that focus the sounds to make them louder. The mathematical model that describes the sound waves can be relatively easy or difficult depending on the situation. If the environment is simple enough, it is possible to estimate with a high degree of certainty the wave equation, but in other situations the equation can become too difficult to analyze.

Frequency, measured in Hertz (Hz), defines the number of complete cycles of a sound wave that occur per second. It is what our ears perceive as pitch. A high-frequency sound has a high pitch (like a piccolo), while a low-frequency sound has a low pitch (like a tuba). The human ear can typically hear frequencies ranging from 20 Hz to 20,000 Hz, although this range decreases with age and exposure to loud noises. Different animals have different hearing ranges; for example, dogs can hear much higher frequencies than humans, which is why dog whistles are silent to our ears but audible to them. Frequencies in the range of 20-20,000 Hz can be subdivided into several categories, namely infrasound, audible sound and ultrasound. The sound above 20,000 Hz is considered ultrasound, and many animals can hear it. Sounds that are below 20 Hz are called infrasound. The lower the frequency, the more energy the sound wave carries.

Amplitude, on the other hand, represents the magnitude of the pressure variations in the sound wave. It is what our ears perceive as loudness. A high-amplitude sound is loud, while a low-amplitude sound is quiet. Amplitude is typically measured in decibels (dB), a logarithmic unit that relates the sound pressure level to a reference pressure. The decibel scale is logarithmic because the human ear is sensitive to a very wide range of sound pressures. A small change in decibels can represent a significant change in sound pressure. For instance, an increase of 10 dB corresponds to a tenfold increase in sound power. To illustrate further, a whisper might be around 30 dB, a normal conversation around 60 dB, and a rock concert around 120 dB. Prolonged exposure to sounds above 85 dB can cause hearing damage. Amplitude and frequency are independent properties of a sound wave. This means that a sound can be high-pitched and quiet, low-pitched and loud, or any combination thereof. Together, they determine the overall characteristic of the sound we perceive. The human ear can detect frequencies in the range of 20 Hz to 20 kHz, and sound pressure from 0 dB to 140 dB. These two parameters define if a sound can be perceived, as well as its individual characteristics, for example tone, volume, and timbre.

The interplay between frequency and amplitude is what gives sound its richness and complexity. A single sound wave with a single frequency and amplitude is a pure tone, like the sound produced by a tuning fork. However, most sounds we encounter in the real world are complex sounds, composed of multiple frequencies and amplitudes. For example, the sound of a musical instrument is made up of a fundamental frequency (the note being played) and a series of overtones or harmonics, which are multiples of the fundamental frequency. The relative amplitudes of these harmonics determine the timbre or tonal quality of the instrument. This is what allows us to distinguish between a violin and a trumpet playing the same note, even though they have the same fundamental frequency. Sound processing techniques are capable of separating the individual frequencies from any sound and analyzing their respective amplitudes to determine the various components that comprise the sound. This process is called spectral analysis, and it is useful for classifying the characteristics of the sounds. For example, one can tell the difference between the sound of a car and a truck by using spectral analysis. The spectral composition can also be represented visually to facilitate its interpretation. This visual representation is called spectrogram.

1.1 Sound Wave Basics

Think of air as a vast ocean; when something vibrates within it, it creates ripples, much like a stone thrown into a still pond. These ripples, these compressions and rarefactions traveling through the air, are what we perceive as sound waves. Understanding their nature is crucial to comprehending not only how sound reaches our ears but also how it can be captured and transmitted remotely. It all begins with a disturbance – perhaps the striking of a guitar string, a shout across a room, or the subtle hum of machinery. These disturbances displace air molecules, initiating a chain reaction where these molecules collide with their neighbors, transferring energy and creating areas of high pressure (compression) and low pressure (rarefaction). These alternating zones of pressure propagate outward from the source, forming the sound wave itself. The speed at which these waves travel depends on the medium through which they move. In air at room temperature, sound travels at approximately 343 meters per second, significantly slower than light. This difference in speed is why we often see lightning before we hear the thunder.

Consider a simple experiment: imagine a loudspeaker cone moving back and forth. As it moves forward, it compresses the air in front of it, creating a region of higher pressure. As it moves backward, it creates a region of lower pressure, or rarefaction. These alternating compressions and rarefactions form a longitudinal wave, meaning that the particles of the medium (air) vibrate parallel to the direction the wave is traveling. It is important to visualize the sound wave not as a movement of air itself – the air molecules don't travel across the room – but as a transfer of energy from one molecule to the next. Each molecule vibrates around its equilibrium position, passing the disturbance along. This is why you can hear sound even if there's a breeze blowing in the opposite direction; the wind affects the movement of air, but it doesn't stop the propagation of the sound wave itself. This concept becomes critical when dealing with environmental factors affecting sound capture. The surrounding elements can either improve or worsen the capture or processing of the wave.

The propagation of sound waves is also influenced by the characteristics of the environment they travel through. Obstacles, such as walls or furniture, can reflect, refract, or absorb sound waves. Reflection occurs when a sound wave bounces off a surface, creating echoes. The angle of incidence (the angle at which the wave hits the surface) is equal to the angle of reflection. Refraction is the bending of sound waves as they pass from one medium to another, or through regions of differing temperature or density. For example, sound waves can bend downward on a cool day because the air near the ground is cooler and denser than the air higher up. This bending can allow sound to travel further than it normally would. Absorption, on the other hand, is the process by which sound energy is converted into other forms of energy, such as heat. Soft, porous materials like carpets and curtains are good absorbers of sound, which is why they are often used in soundproofing. All of these phenomena – reflection, refraction, and absorption – play a crucial role in how we perceive sound in different environments, and they are important considerations in remote listening applications.

Furthermore, the behavior of sound waves is governed by the principle of superposition. This principle states that when two or more sound waves meet at a point, the resulting displacement at that point is the sum of the displacements of the individual waves. This leads to phenomena like constructive and destructive interference. Constructive interference occurs when the crests of two waves coincide, resulting in a wave with a larger amplitude (and thus a louder sound). Destructive interference occurs when the crest of one wave coincides with the trough of another, resulting in a wave with a smaller amplitude (and potentially silence if the waves are of equal amplitude and opposite phase). Understanding interference is crucial in designing microphone arrays and other signal processing techniques used in remote listening. For instance, noise-canceling headphones use destructive interference to eliminate unwanted background noise. By generating a sound wave that is the inverse of the ambient noise, they effectively cancel it out, allowing the listener to hear only the desired sound.

Moving beyond the fundamentals, it's important to recognize that sound waves don't travel indefinitely. They lose energy as they propagate through a medium, a phenomenon known as attenuation. Attenuation is caused by several factors, including spreading, absorption, and scattering. Spreading refers to the decrease in sound intensity as the wave expands outward from the source. In a free field (an environment with no reflections), the sound intensity decreases proportionally to the square of the distance from the source. Absorption, as mentioned earlier, is the conversion of sound energy into other forms of energy. Scattering occurs when sound waves encounter obstacles that are smaller than their wavelength, causing the waves to be redirected in various directions. Attenuation is a critical consideration in remote listening applications, as it limits the range over which sound can be effectively captured and transmitted. Understanding these wave behaviors helps to optimize the settings and equipment needed for sound perception.

Lastly, to really drive home this point, think about how we use these basics every day without even thinking. Imagine you are at a concert. The sound waves are generated by the instruments on stage, travel through the air, and reach your ears. The acoustics of the venue – the size and shape of the room, the materials used in its construction – affect how the sound waves are reflected, refracted, and absorbed. This affects the overall sound quality you experience. Or consider a phone call. Your voice is converted into an electrical signal, transmitted over a network, and then converted back into sound waves at the other end. The quality of the call depends on the fidelity of these conversions and the characteristics of the transmission medium. Understanding the basics of sound waves provides a foundation for understanding these and countless other everyday experiences, and it is absolutely essential for anyone working with sound, whether in music, acoustics, engineering, or any other field. It's about more than just hearing; it's about understanding the physics of what we hear and how we can manipulate it.

1.2 Frequency and Amplitude

The journey into understanding sound begins with its most fundamental characteristics: frequency and amplitude. These two properties, while seemingly simple, govern the essence of how we perceive the world through the auditory sense. It's like understanding the brushstrokes and the colors before appreciating a masterpiece; frequency and amplitude are the building blocks upon which all other acoustic phenomena are built. One cannot fully grasp the complexities of remote listening without first mastering these core concepts.

Imagine plucking a guitar string. The string vibrates, creating a disturbance in the air around it. This disturbance propagates outwards as a sound wave, a pattern of alternating compressions and rarefactions of air molecules. Frequency, in its simplest definition, is the rate at which these compressions and rarefactions occur. It's the number of complete cycles of the wave that pass a given point in one second. We measure frequency in Hertz (Hz), where 1 Hz signifies one cycle per second. A higher frequency means more cycles per second, which our ears perceive as a higher pitch. Think of a piccolo, with its high, piercing notes – these are sounds with high frequencies. Conversely, a tuba produces low, rumbling sounds with low frequencies. The human ear can typically detect sounds ranging from approximately 20 Hz to 20,000 Hz, although this range diminishes with age. Understanding the frequency range of different sound sources is crucial in remote listening, particularly when targeting specific sounds amidst background noise. For instance, monitoring wildlife might involve focusing on the specific frequency ranges of bird calls or animal vocalizations, filtering out irrelevant sounds.

Amplitude, on the other hand, represents the intensity or power of the sound wave. It's the magnitude of the pressure variations in the air caused by the sound wave. A sound wave with a large amplitude carries more energy and creates a greater displacement of air molecules. We perceive amplitude as loudness. A shout has a much higher amplitude than a whisper, and a jackhammer produces sounds with a significantly higher amplitude than a ticking clock. Amplitude is typically measured in decibels (dB), a logarithmic scale that allows us to represent a wide range of sound intensities in a manageable way. The decibel scale is crucial because our ears perceive loudness logarithmically rather than linearly. An increase of 10 dB corresponds to roughly a doubling of perceived loudness. The quietest sound a human ear can typically detect is defined as 0 dB. A sound at 100 dB is ten billion times more intense than a sound at 0 dB, highlighting the vast range that the decibel scale encompasses. In remote listening applications, understanding amplitude is essential for detecting faint sounds at a distance. A microphone's sensitivity and the amplification of the signal play key roles in capturing and processing these low-amplitude sounds.

The relationship between frequency and amplitude is not always straightforward. While they are independent properties, they often interact in complex ways. For example, the human ear is not equally sensitive to all frequencies at the same amplitude. We are most sensitive to frequencies in the range of 1 kHz to 4 kHz, which is the range of human speech. This means that a sound at 1 kHz will sound louder than a sound at 100 Hz, even if they have the same amplitude. This non-linear response of the human ear is a crucial consideration in psychoacoustics, the study of how we perceive sound.

Consider the challenge of recording a conversation in a crowded room. The room is filled with sounds of varying frequencies and amplitudes: voices, music, clinking glasses, and the hum of air conditioning. To isolate the conversation of interest, a remote listening system would need to selectively amplify the frequencies corresponding to human speech while attenuating the frequencies of the background noise. This process often involves filtering, a technique that selectively blocks or passes certain frequency ranges. Furthermore, the system would need to differentiate between the quiet voices of the speakers and the louder sounds of the environment. This requires careful adjustment of the microphone's gain and the dynamic range of the recording, the difference between the quietest and loudest sounds that the system can capture without distortion.

In wildlife monitoring, the task might involve detecting the subtle calls of specific bird species in a forest teeming with sounds of wind, rustling leaves, and other animal vocalizations. By knowing the specific frequency range of the target bird's call, researchers can use specialized recording equipment to filter out the extraneous noises and isolate the desired signal. The amplitude of the bird's call can also provide information about its distance and activity level. A faint call might indicate that the bird is far away, while a loud call might suggest that it is nearby or actively communicating. Analyzing both the frequency and amplitude characteristics of the sound can reveal crucial insights into the bird's behavior and population dynamics.

Moreover, environmental factors can significantly affect both frequency and amplitude. As a sound wave travels through the air, it can be absorbed, reflected, or refracted by various objects and atmospheric conditions. High-frequency sounds tend to be absorbed more readily than low-frequency sounds, which is why distant sounds often sound muffled. Amplitude also decreases with distance due to the spreading of the sound wave and energy loss due to absorption. Understanding these environmental influences is crucial for interpreting remote listening data accurately. Acoustic modeling tools can be used to predict how sound waves will propagate in a given environment, taking into account factors such as temperature, humidity, and terrain.

As technology evolves, the ability to manipulate and analyze frequency and amplitude has become increasingly sophisticated. Digital signal processing techniques allow us to precisely control the frequency content of audio signals, creating new and innovative applications in fields such as music production, telecommunications, and medical diagnostics. For instance, in medical applications, remote listening devices can be used to monitor heart and lung sounds, detecting subtle changes in frequency and amplitude that may indicate underlying health problems. These devices can transmit the audio data wirelessly to a remote location, allowing physicians to monitor patients from afar. In security, analyzing the specific frequencies within recorded audio can detect the unique patterns of certain events, potentially distinguishing the sound of breaking glass from other ambient noises.

Mastering the concepts of frequency and amplitude is not merely an academic exercise; it's a foundational skill that empowers us to understand and interact with the world around us in profound ways. From the subtle nuances of a musical performance to the critical information conveyed by a distant warning signal, frequency and amplitude are the keys to unlocking the hidden language of sound. By understanding these principles, we can develop increasingly sophisticated remote listening systems that enhance our ability to perceive and interpret the acoustic environment, opening up new possibilities in science, technology, and human understanding.

2 Chapter 2: Transduction Mechanisms

Microphones serve as the crucial first step in the transduction process, acting as the interface between the acoustic world and the electronic domain. They perform the essential task of converting sound waves, which are essentially variations in air pressure, into corresponding electrical signals. This conversion is achieved through a variety of mechanisms, each with its own set of advantages and disadvantages, making them suitable for different applications and acoustic environments. Dynamic microphones, known for their robustness and ability to handle high sound pressure levels, operate based on the principle of electromagnetic induction. A diaphragm, typically made of a thin plastic or metal film, is attached to a coil of wire suspended within a magnetic field. When sound waves strike the diaphragm, it vibrates, causing the coil to move within the magnetic field. This movement induces a voltage in the coil, proportional to the velocity of the diaphragm's movement and thus, to the amplitude and frequency of the sound wave. Due to their relatively simple construction, dynamic microphones are durable and can withstand harsh conditions, making them a popular choice for live performances and field recording where reliability is paramount. Condenser microphones, on the other hand, employ a different principle based on capacitance. They consist of two plates, one fixed and the other movable, forming a capacitor. The movable plate, acting as the diaphragm, vibrates in response to sound waves, causing the distance between the plates to change. This change in distance alters the capacitance of the capacitor. A voltage is applied across the capacitor, and the variations in capacitance are converted into an electrical signal. Condenser microphones are typically more sensitive and have a wider frequency response than dynamic microphones, allowing them to capture more subtle nuances in sound. However, they require a power source, either from an external power supply (phantom power) or an internal battery, to operate.

Following the initial capture by the microphone, the electrical signal representing the sound wave undergoes a series of crucial conversion stages designed to optimize it for transmission, storage, or further processing. Amplification is often the first step, increasing the signal's voltage or current to a level that is suitable for subsequent stages. This process ensures that the signal is strong enough to overcome noise and interference that may be present in the system. The type of amplifier used depends on the characteristics of the signal and the requirements of the application, with considerations such as gain, bandwidth, and noise figure being carefully chosen. Filtering is another essential stage, used to selectively attenuate or remove unwanted frequencies from the signal. This can include removing low-frequency rumble, high-frequency hiss, or specific interfering signals. Filters are typically implemented using electronic circuits, such as active filters or passive filters, designed to have specific frequency response characteristics. The choice of filter type and parameters depends on the specific application and the characteristics of the noise or interference that needs to be removed. Following amplification and filtering, the signal may undergo additional processing stages, such as equalization, compression, or expansion, depending on the desired outcome. Equalization is used to adjust the frequency balance of the signal, enhancing certain frequencies or attenuating others. Compression is used to reduce the dynamic range of the signal, making it louder and more consistent. Expansion is the opposite of compression, increasing the dynamic range of the signal. These processing stages can be implemented using analog circuits or digital signal processing (DSP) techniques, offering flexibility and control over the final sound.

The culmination of the transduction process involves the conversion of the analog electrical signal into a digital representation, a process known as analog-to-digital conversion (ADC). This conversion is essential for modern audio systems, as it allows the signal to be stored, transmitted, and processed digitally. The ADC samples the analog signal at discrete points in time and quantizes each sample to a specific digital value. The sampling rate, measured in samples per second (Hz), determines how often the signal is sampled. A higher sampling rate captures more of the signal's information, resulting in a more accurate digital representation. The Nyquist-Shannon sampling theorem states that the sampling rate must be at least twice the highest frequency present in the analog signal to avoid aliasing, a phenomenon where high-frequency components are misrepresented as lower-frequency components. Common sampling rates used in audio applications include 44.1 kHz (CD quality) and 48 kHz (professional audio). The bit depth, also known as resolution, determines the number of bits used to represent each sample. A higher bit depth allows for more precise quantization, resulting in a wider dynamic range and lower quantization noise. Common bit depths used in audio applications include 16 bits (CD quality) and 24 bits (professional audio). The digital representation of the audio signal is then encoded using a specific audio codec, such as PCM (Pulse Code Modulation), MP3 (MPEG-1 Audio Layer 3), or AAC (Advanced Audio Coding). Each codec uses different compression techniques to reduce the file size of the audio signal, while preserving as much of the original sound quality as possible. The choice of codec depends on the specific application and the desired trade-off between file size and sound quality. The resulting digital audio data can then be stored on a digital storage medium, transmitted over a digital network, or processed using digital signal processing techniques. Digital representation allows for greater flexibility, ease of manipulation, and robustness against noise and distortion compared to analog signals.

2.1 Microphone Principles

Sound is vibration, and microphones are the transducers that convert these vibrations into electrical signals we can then amplify, record, or transmit. Understanding the inner workings of these devices is crucial for anyone working with audio, from recording a simple voice memo to designing a sophisticated surveillance system. Without a grasp of microphone principles, you're effectively flying blind, hoping for the best without knowing how your equipment actually operates. It's a little deeper than "point and record."

The fundamental principle uniting all microphones is transduction. This is the process of converting one form of energy – in this case, acoustic energy (sound waves) – into another form of energy, electrical energy (an electrical signal). This electrical signal, ideally, is a faithful representation of the original sound. The better the microphone, the more accurately it captures the nuances of the sound wave. Many technologies do this, but this chapter will focus on the major ones.

Dynamic microphones are rugged workhorses, commonly found on stage and in recording studios. Their transduction mechanism relies on electromagnetic induction. A thin diaphragm, typically made of Mylar or a similar material, is attached to a coil of wire. This coil is suspended within the magnetic field of a permanent magnet. When sound waves strike the diaphragm, it vibrates, causing the coil to move within the magnetic field. This movement induces a voltage in the coil, proportional to the velocity of the diaphragm's movement. This voltage is the electrical signal representing the sound. Dynamic microphones are known for their ability to handle high sound pressure levels (SPL), making them suitable for loud sources like drums or guitar amplifiers. They don't require external power, which adds to their versatility. The construction makes them tough and resilient to the rigors of live performance, making them a reliable option.

Condenser microphones operate on a different principle: variable capacitance. A condenser microphone consists of two plates, one fixed and one movable (the diaphragm), forming a capacitor. A capacitor stores electrical energy. The diaphragm is typically made of a thin, metallized plastic film. When sound waves strike the diaphragm, it vibrates, changing the distance between the two plates and, consequently, changing the capacitance. This change in capacitance is converted into a voltage signal. Because the change in capacitance is very small, condenser microphones require an external power source to operate. This power, often referred to as phantom power (+48V), is typically supplied by the recording console or audio interface. Condenser microphones are generally more sensitive than dynamic microphones and offer a wider frequency response, capturing more detail and nuance in the sound. This makes them ideal for recording vocals, acoustic instruments, and other sources where detail is important.

Electret condenser microphones are a variation of the condenser type, but with a key difference: they use a permanently charged material (the electret) to create the electric field. This eliminates the need for an external polarizing voltage, although they still typically require power for the internal impedance converter (often a battery or phantom power). They are commonly found in smartphones, laptops, and lavalier microphones due to their small size and relatively low cost. While not always offering the same level of performance as true condenser microphones, electret condensers provide a decent balance of size, cost, and performance for many applications. The tradeoff is their lower dynamic range and potential for self-noise.

Ribbon microphones are a more specialized type of microphone that utilizes a thin, corrugated ribbon of metal suspended within a magnetic field. When sound waves strike the ribbon, it vibrates, inducing a voltage in the ribbon due to electromagnetic induction. Ribbon microphones are known for their warm, smooth sound and their characteristic figure-8 polar pattern (equally sensitive to sound from the front and back, but rejects sound from the sides). They are often used on vocals, brass instruments, and guitar amplifiers to add a vintage, "rounded" tone. However, ribbon microphones are generally more fragile than dynamic microphones and require careful handling. They are also typically less sensitive, requiring more gain from the preamplifier. They can be easily damaged by strong air blasts or phantom power, so you must be cautious when using them.

Piezoelectric microphones, also known as crystal microphones, utilize the piezoelectric effect. This effect occurs in certain materials (like quartz crystals) that generate an electrical charge when subjected to mechanical stress. In a piezoelectric microphone, a crystal is attached to a diaphragm. When sound waves vibrate the diaphragm, it applies stress to the crystal, generating a voltage. Piezoelectric microphones are typically small and inexpensive, but their sound quality is generally not as high as other types of microphones. They are often used in contact microphones for acoustic instruments, picking up vibrations directly from the instrument's body. Their low cost makes them good for basic applications.

Optical microphones represent a newer and less common technology. They use light to detect sound vibrations. Typically, a light source shines onto a reflective diaphragm. The movement of the diaphragm modulates the reflected light, which is then detected by a photosensor. The changes in the light signal are converted into an electrical signal. Optical microphones offer several potential advantages, including high sensitivity, low noise, and wide frequency response. They are also immune to electromagnetic interference. However, they are typically more expensive and complex than other types of microphones.

The choice of microphone technology depends heavily on the specific application. For live sound reinforcement where robustness is key, dynamic microphones are often preferred. For critical studio recording where detail is paramount, condenser microphones are usually the better choice. Ribbon microphones can add a unique sonic character, while electret condensers offer a compact and affordable solution for various applications. Optical microphones are still emerging, but they hold promise for high-performance audio capture in demanding environments. Careful consideration of the microphone's transduction mechanism, polar pattern, frequency response, and other specifications is essential for achieving the desired results. Each technology has its strengths and weaknesses, and understanding these nuances is crucial for making informed decisions about microphone selection and usage. Remember, the right microphone can make all the difference in the quality of your audio recordings and transmissions.

2.2 Signal Conversion Stages

The journey of a sound wave, initially a simple vibration, transforms into a complex electrical representation through a series of crucial stages. Each stage plays a vital role in preparing the signal for efficient transmission, processing, and ultimately, faithful reproduction. This process isn't merely about conversion; it's about optimizing the signal for clarity and minimizing the introduction of unwanted artifacts. Initially, the acoustic energy must be converted into an electrical signal by a transducer, typically a microphone. This initial conversion is crucial because the quality of the final audio is heavily dependent on the sensitivity and accuracy of the microphone. The diaphragm of the microphone vibrates in response to sound waves, and this mechanical movement is then converted into a corresponding electrical signal. Different microphone technologies, such as dynamic, condenser, and electret microphones, employ varied mechanisms to achieve this conversion, each with its own set of advantages and disadvantages in terms of frequency response, sensitivity, and noise floor. The electrical signal produced at this stage is usually very weak and needs to be amplified to a level suitable for further processing.

Once the fragile electrical signal emerges from the microphone, amplification becomes paramount. This critical stage increases the signal's magnitude, making it robust enough for subsequent manipulation. Pre-amplifiers, specifically designed for this purpose, are employed to boost the signal without introducing excessive noise. The choice of pre-amplifier is crucial, as it directly affects the signal-to-noise ratio (SNR) of the entire system. A good pre-amplifier will add minimal noise to the signal while providing sufficient gain to overcome the noise floor of subsequent stages. Amplification isn't a singular, linear process; it often involves multiple stages, each carefully calibrated to avoid clipping or distortion. Over-amplification can lead to saturation, where the signal is forced beyond the amplifier's capacity, resulting in a flattened waveform and an unpleasant distorted sound. Conversely, under-amplification leaves the signal vulnerable to interference and noise, compromising its clarity. Feedback loops are frequently incorporated into pre-amplifier designs to stabilize gain and minimize distortion. Equalization may also be applied at this stage to tailor the frequency response of the signal, compensating for the microphone's characteristics or the acoustic properties of the environment.

Following amplification, the signal often undergoes filtering to remove unwanted noise and shape the frequency spectrum. Filtering is an essential process that sculpts the audio signal, removing frequencies that are not relevant or that contribute to noise. High-pass filters, for instance, attenuate low-frequency rumble and hum, while low-pass filters remove high-frequency hiss and aliasing artifacts. Band-pass filters can isolate specific frequency ranges, useful for emphasizing particular instruments or vocal frequencies. The design of filters involves considerations of filter order, cutoff frequency, and filter type (e.g., Butterworth, Bessel, Chebyshev). Higher-order filters provide steeper attenuation slopes, but they can also introduce phase distortion. The cutoff frequency determines the point at which the filter begins to attenuate the signal, and it must be carefully chosen to avoid removing essential audio information. Notch filters can be used to remove specific frequencies, such as mains hum (50 or 60 Hz), without affecting the surrounding frequencies. Dynamic filters can adjust their parameters in response to the input signal, providing adaptive noise reduction. For example, a noise gate can automatically attenuate the signal when it falls below a certain threshold, effectively silencing background noise during pauses in speech.

After the meticulous shaping and cleaning of the signal through filtering, the stage is set for encoding, the process of converting the analog signal into a digital representation. Encoding transforms the continuous analog signal into a discrete digital format, ready for storage, transmission, or digital processing. This process involves two key steps: sampling and quantization. Sampling captures the amplitude of the analog signal at discrete points in time, determined by the sampling rate. The Nyquist-Shannon sampling theorem dictates that the sampling rate must be at least twice the highest frequency present in the signal to avoid aliasing. Common sampling rates include 44.1 kHz (CD quality) and 48 kHz (professional audio). Quantization assigns a discrete value to each sample, representing its amplitude. The number of bits used for quantization determines the dynamic range of the digital signal. For example, 16-bit quantization provides a dynamic range of 96 dB, while 24-bit quantization offers 144 dB. Quantization introduces quantization noise, which can be reduced by using dithering, a technique that adds a small amount of random noise to the signal. The encoded digital signal can then be compressed using various audio codecs, such as MP3, AAC, or FLAC, to reduce the file size for efficient storage and transmission. The choice of codec depends on the desired balance between file size and audio quality. Lossless codecs, such as FLAC, preserve all the original audio information, while lossy codecs, such as MP3, discard some information to achieve greater compression.

The transmission stage is a crucial component, especially when dealing with remote listening applications. Wireless communication and networked systems demand careful consideration of bandwidth, latency, and reliability. Whether employing radio frequencies, Bluetooth, Wi-Fi, or wired networks, the transmission channel introduces its own set of challenges. Bandwidth limitations restrict the amount of data that can be transmitted per unit of time, influencing the choice of audio codec and compression settings. Latency, the delay between signal capture and reception, can be particularly problematic in interactive applications, such as remote conferencing or live music performances. Reliable transmission is essential to prevent dropouts and errors, which can significantly degrade the audio quality. Error correction techniques, such as forward error correction (FEC), can be used to mitigate the effects of data loss during transmission. Networked audio systems often employ protocols such as RTP (Real-time Transport Protocol) to ensure reliable and timely delivery of audio packets. The selection of the appropriate transmission technology and protocols depends on the specific application requirements and the available infrastructure.

Finally, the received signal undergoes decoding and playback to recreate the original sound. The decoding process reverses the encoding process, converting the digital signal back into an analog signal. This involves decompressing the audio data, performing digital-to-analog conversion (DAC), and amplifying the signal to a level suitable for driving headphones or speakers. The quality of the DAC is critical to the fidelity of the reproduced audio. High-quality DACs minimize distortion and noise, providing a cleaner and more accurate representation of the original signal. The playback stage also involves equalization and other signal processing techniques to optimize the audio for the specific listening environment. Room acoustics can significantly affect the perceived sound quality, and equalization can be used to compensate for these effects. Loudness normalization techniques can be applied to ensure that the audio is played back at a consistent volume level, regardless of the source material. The ultimate goal of the playback stage is to deliver an immersive and enjoyable listening experience that accurately reflects the original sound. Each of these stages, from the initial acoustic transduction to the final playback, is carefully orchestrated to ensure that the audio signal is faithfully captured, processed, and reproduced.

2.3 Digital Representation

Once sound has been captured and transduced into an electrical signal by a microphone, the next crucial step involves converting that analog signal into a digital representation. This process, called analog-to-digital conversion (ADC), is fundamental to any remote listening system that relies on digital transmission, storage, or processing. The accuracy and efficiency of this conversion directly affect the quality of the audio signal received at the remote end, influencing the listener's perception and interpretation of the sound event. Understanding the nuances of digital representation is therefore paramount.

Sampling is the first essential procedure in ADC. It involves taking discrete snapshots of the continuous analog waveform at regular intervals. The frequency at which these samples are taken, known as the sampling rate, determines the maximum frequency content that can be accurately represented in the digital domain. According to the Nyquist-Shannon sampling theorem, the sampling rate must be at least twice the highest frequency present in the original signal to avoid aliasing, which is a distortion that occurs when high-frequency components are misinterpreted as lower frequencies. For instance, audio intended for human consumption typically uses a sampling rate of 44.1 kHz, which can accurately capture frequencies up to 22.05 kHz, exceeding the typical range of human hearing. Lower sampling rates may be sufficient for applications where high-frequency content is not critical, such as voice communication, where 8 kHz or 16 kHz sampling rates are commonly used. The chosen sampling rate represents a trade-off between fidelity and data size, as higher sampling rates result in larger data files and increased bandwidth requirements for transmission.

Following sampling, the next crucial step is quantization, in which the amplitude of each sample is assigned a discrete value from a predefined set of levels. The number of quantization levels determines the dynamic range and signal-to-noise ratio of the digital signal. A higher number of levels provides a finer resolution, resulting in a more accurate representation of the original analog signal and reducing quantization noise, which sounds like a low-level hiss. The number of levels is usually expressed in bits, where an n-bit quantizer has 2^n levels. For example, 16-bit audio, commonly used in CDs, has 65,536 levels, while 24-bit audio, often used in professional recording, has over 16 million levels, enabling a wider dynamic range and lower noise floor. The choice of bit depth depends on the specific application and the desired level of audio fidelity. Lower bit depths can be used to reduce data size and bandwidth requirements, but at the expense of increased quantization noise and reduced dynamic range.

The final stage in digital representation involves encoding the quantized samples into a digital format suitable for storage or transmission. Several encoding schemes are available, each with its own advantages and disadvantages in terms of compression efficiency, computational complexity, and audio quality. Pulse Code Modulation (PCM) is the simplest encoding scheme, where each sample is represented by a binary code corresponding to its quantized level. PCM is widely used as a baseline format for uncompressed audio and is the foundation for many other encoding schemes. However, PCM can be inefficient for storing or transmitting audio, as it does not exploit any redundancy in the signal. To address this limitation, compression algorithms are used to reduce the data size while preserving the perceptual quality of the audio. Lossless compression algorithms, such as FLAC (Free Lossless Audio Codec) and ALAC (Apple Lossless Audio Codec), reduce the data size without discarding any information, allowing for perfect reconstruction of the original audio signal. These algorithms exploit statistical redundancies in the audio data to achieve compression ratios of typically 2:1 or 3:1. Lossy compression algorithms, such as MP3 (MPEG-1 Audio Layer III) and AAC (Advanced Audio Coding), achieve higher compression ratios by discarding some information that is deemed perceptually irrelevant. These algorithms use psychoacoustic models to identify and remove frequencies and amplitudes that are unlikely to be noticed by the human ear.

Lossy compression can achieve compression ratios of 10:1 or higher, but at the cost of some degradation in audio quality. The choice of encoding scheme depends on the specific application and the trade-off between data size, bandwidth requirements, and audio quality.

Beyond the fundamental steps of sampling, quantization, and encoding, advanced digital representation techniques can further enhance the quality and efficiency of remote listening systems. For example, noise shaping techniques can be used to redistribute quantization noise to frequencies where it is less audible, thereby improving the perceived signal-to-noise ratio. Dithering, adding a small amount of random noise to the analog signal before quantization, can reduce the audibility of quantization distortion, especially for low-level signals. Differential coding techniques, such as Delta Modulation and Adaptive Differential Pulse Code Modulation (ADPCM), encode the difference between successive samples rather than the absolute values, which can be more efficient for signals with high correlation between samples. These techniques can reduce the data rate while preserving the essential information in the audio signal. Multichannel audio encoding techniques, such as Dolby Digital and DTS, are used to represent surround sound audio, enabling immersive and realistic remote listening experiences. These techniques encode multiple audio channels, corresponding to different spatial locations, and transmit them over a single digital stream. At the receiving end, the audio channels are decoded and played back through a multi-speaker system, creating a spatial audio image that replicates the original sound field.

Furthermore, the selection of appropriate hardware components for analog-to-digital conversion is crucial for achieving high-quality digital representation. The ADC chip itself should have sufficient resolution and accuracy to meet the requirements of the application. Factors such as the sampling rate, bit depth, and signal-to-noise ratio of the ADC should be carefully considered. Additionally, the analog front-end circuitry, including amplifiers, filters, and impedance matching networks, should be designed to minimize noise and distortion and to optimize the signal level for the ADC. The clock source used to control the sampling rate should be stable and accurate to avoid jitter, which can degrade the audio quality. Finally, the power supply should be clean and well-regulated to prevent noise from entering the analog signal path. By carefully selecting and designing the hardware components, it is possible to achieve a high-quality digital representation of the audio signal that accurately captures the nuances and details of the original sound event. These choices affect the quality and power of the digital signal being transmitted.

In essence, mastering digital representation is not just about understanding the technical specifications of ADCs and codecs; it's about making informed decisions that balance fidelity, efficiency, and the specific needs of the remote listening application. Whether it's optimizing sampling rates for clear speech transmission or choosing the right compression algorithm for archival purposes, a solid grasp of these principles ensures that the sound captured is the sound perceived, no matter the distance. The proper digital representation has a big impact in remote listening applications.

3 Chapter 3: Remote Listening Technologies

Remote listening technologies hinge on the ability to capture, transmit, and reconstruct acoustic information across spatial distances. This chapter dives into the core technologies enabling this functionality, focusing on the two primary mechanisms: wireless transmission and networked audio systems. Understanding these methodologies is crucial for anyone seeking to implement or analyze remote audio monitoring or surveillance solutions. We'll look closely at these technologies.

Wireless transmission, in its essence, uses electromagnetic waves to carry audio signals through the air. Radio frequency (RF) transmission remains a prevalent technique, utilizing various frequency bands to modulate and broadcast audio. Common examples include FM radio, which, while primarily used for broadcast entertainment, can be adapted for dedicated monitoring applications. More relevant are technologies like Bluetooth and Wi-Fi, which offer digital transmission, thus superior audio quality and security features compared to traditional analog RF systems. Bluetooth, with its low power consumption and short-range capabilities, is well-suited for applications like wireless microphones and personal listening devices. Wi-Fi, on the other hand, provides a longer range and higher bandwidth, enabling the transmission of high-fidelity audio streams over larger areas. The choice of wireless technology depends heavily on the specific application requirements, including factors like range, bandwidth, power consumption, and security. Beyond these established standards, specialized wireless protocols like Zigbee and Z-

Wave are emerging, particularly in the context of smart home and industrial monitoring systems, providing robust, low-power communication for sensor networks. These networks can include acoustic sensors, enabling remote audio capture in environments where traditional Wi-Fi or Bluetooth may be less suitable due to range or power limitations. Careful consideration must be given to antenna design and placement to optimize signal strength and minimize interference, especially in complex environments. We must consider the environment too.
Networked audio systems leverage digital networks, primarily the internet, to transport audio signals. This approach enables global reach, facilitating remote listening from virtually anywhere with an internet connection. The foundation of networked audio lies in audio codecs, which compress audio signals for efficient transmission over the network. Codecs like MP3, AAC, and Opus offer varying degrees of compression and audio quality. Choosing the right codec is crucial for balancing bandwidth requirements with the desired level of fidelity. Protocols like RTP (Real-time Transport Protocol) and SIP (Session Initiation Protocol) are commonly used to manage audio streams over IP networks, ensuring reliable delivery and synchronization. RTP provides mechanisms for packet sequencing and error correction, while SIP handles session initiation and call control. Cloud-based platforms offer a centralized infrastructure for managing and distributing audio streams, enabling scalable and accessible remote listening solutions. These platforms often provide features like user authentication, access control, and audio recording, enhancing the security and management of remote audio monitoring systems. Network latency, the delay in transmitting data over the network, can significantly impact the perceived quality of real-time audio streams. Techniques like buffering and jitter compensation are employed to mitigate the effects of latency, but careful network planning and optimization are essential for minimizing delays. Security is paramount in networked audio systems, as audio streams may contain sensitive information. Encryption protocols like TLS/SSL should be used to protect audio streams from eavesdropping. Proper authentication and authorization mechanisms are crucial for ensuring that only authorized users can access the audio streams. We need to be really secure.

While these technologies offer immense capabilities, it's essential to acknowledge their limitations. Wireless transmission is susceptible to interference from other electromagnetic sources, leading to signal degradation or loss. Networked audio systems are vulnerable to network congestion and latency, which can disrupt audio streams. Furthermore, the security of both wireless and networked systems is a constant concern, requiring vigilant monitoring and proactive security measures. Ethical considerations surrounding the use of these technologies are also paramount. Remote listening capabilities must be deployed responsibly, with a strong emphasis on protecting individual privacy and adhering to legal regulations. The potential for misuse, such as unauthorized surveillance, must be carefully considered and mitigated. Only by addressing these technical and ethical challenges can we ensure that remote listening technologies are used in a safe, responsible, and beneficial manner. There are many issues that need to be taken into account.

Moreover, the implementation of remote listening technologies often involves integrating different components into a cohesive system. For example, a remote wildlife monitoring system might incorporate acoustic sensors, wireless transmitters, and cloud-based data storage. This integration requires careful planning and testing to ensure that all components work together seamlessly. Signal processing techniques, such as noise reduction and echo cancellation, can be applied to enhance the quality of the captured audio, improving the accuracy and reliability of remote monitoring applications. Proper calibration of acoustic sensors is crucial for ensuring accurate measurements of sound levels and frequencies. Regular maintenance and monitoring of the system are essential for identifying and addressing any potential issues, such as sensor drift or network connectivity problems. The scalability of the system should also be considered, particularly for applications that require monitoring multiple locations simultaneously. Designing a scalable system involves using appropriate hardware and software architectures that can handle the increasing volume of data and user requests.

Ultimately, the successful deployment of remote listening technologies requires a multidisciplinary approach, combining expertise in acoustics, signal processing, wireless communications, and network security. By carefully considering the technical challenges, ethical implications, and integration requirements, we can harness the power of these technologies to create innovative solutions for a wide range of applications.

3.1 Wireless Transmission

Wireless transmission forms the backbone of many remote listening applications, enabling the capture and relay of audio signals without physical cables. This capability is paramount for scenarios where wired connections are impractical, such as wildlife monitoring in remote locations, discreet surveillance operations, or assistive listening devices used in public spaces. Understanding the nuances of wireless transmission is crucial for designing robust and reliable remote listening systems.

The foundation of wireless audio lies in converting sound into electromagnetic waves, which propagate through the air. At the simplest level, amplitude modulation (AM) and frequency modulation (FM) techniques can be employed. AM varies the amplitude of a carrier wave to match the amplitude of the audio signal, while FM alters the carrier wave's frequency. FM generally offers superior sound quality and noise immunity compared to AM, making it a preferred choice for many applications. However, both AM and FM have limitations in terms of bandwidth and susceptibility to interference. The effective use of each protocol depends on the context in which they are implemented and also the expected performance from them.

Advancements in digital technology have paved the way for more sophisticated wireless transmission methods. Bluetooth, Wi-Fi, and cellular networks are now commonly used for transmitting audio signals over varying distances. Bluetooth, a short-range wireless technology, is well-suited for personal audio devices like headphones and speakers. Its low power consumption and ease of pairing make it a convenient option for localized audio transmission. Wi-Fi, on the other hand, offers higher bandwidth and longer range, making it suitable for streaming audio over local area networks. This allows for multi-room audio systems, networked surveillance systems, and other applications that require wider coverage. Cellular networks provide the broadest coverage, enabling audio transmission across vast geographical distances. This is particularly useful for applications like remote monitoring and long-distance communication.

Several factors influence the performance of wireless audio transmission. Signal strength is paramount; weak signals can result in dropouts, distortion, and reduced audio quality. Distance, obstacles, and interference from other electronic devices can all attenuate the signal. Careful site planning and antenna placement can help mitigate these issues. Proper antenna design and positioning can significantly improve signal strength and coverage, particularly in challenging environments. Also, selecting the appropriate wireless technology for the specific application is crucial. Bluetooth is adequate for short-range, low-bandwidth applications, while Wi-Fi or cellular networks may be necessary for longer distances or higher data rates.

Bandwidth also plays a significant role in audio quality. Higher bandwidth allows for the transmission of a wider range of frequencies, resulting in richer, more detailed audio. Lossy compression algorithms like MP3 and AAC are commonly used to reduce the file size of audio signals, enabling efficient transmission over wireless networks. However, compression can introduce artifacts and reduce audio quality. The choice of compression algorithm and bit rate represents a trade-off between file size and audio fidelity. For critical applications, lossless compression algorithms like FLAC or uncompressed formats like WAV may be preferred, although they require significantly more bandwidth.

Security is another critical consideration in wireless audio transmission. Wireless signals are inherently vulnerable to eavesdropping, making it essential to implement appropriate security measures. Encryption is a fundamental technique for protecting audio data from unauthorized access. Protocols like WPA2 and WPA3 provide strong encryption for Wi-Fi networks, while Bluetooth offers built-in encryption capabilities. Furthermore, secure authentication mechanisms can prevent unauthorized devices from connecting to the wireless network. These measures are particularly important in sensitive applications like surveillance and medical diagnostics, where privacy and confidentiality are paramount.

Furthermore, power consumption is a crucial factor in many remote listening applications, especially those operating on battery power. Wireless transmission can be energy-intensive, so optimizing power consumption is essential for extending battery life. Techniques like duty cycling, where the transmitter is only active for short periods, can significantly reduce power consumption. Low-power wireless technologies like Bluetooth Low Energy (BLE) are also available, offering significantly lower power consumption compared to traditional Bluetooth. The choice of wireless technology and power management strategies should be carefully considered based on the specific application requirements.

The selection of appropriate hardware and software is also crucial for successful wireless audio transmission. High-quality microphones and audio interfaces are essential for capturing clean and accurate audio signals. The wireless transmitter and receiver should be carefully selected to match the specific application requirements, considering factors like range, bandwidth, security, and power consumption. Software tools for encoding, decoding, and managing audio streams are also necessary. Open-source software libraries like GStreamer provide a flexible and powerful platform for audio processing. Moreover, understanding the regulatory landscape is vital. Wireless communication is subject to regulations that vary by region. Compliance with these regulations is essential to avoid legal issues and ensure interoperability. It's important to be aware of frequency bands, power limits, and other technical requirements imposed by regulatory bodies.

Real-world implementation often presents unique challenges. Environmental conditions, such as temperature, humidity, and electromagnetic interference, can affect the performance of wireless systems. Careful testing and calibration are essential for ensuring reliable operation in the intended environment. Regular maintenance and updates are also necessary to address security vulnerabilities and improve performance. With careful planning, implementation, and ongoing maintenance, wireless transmission can provide a reliable and cost-effective solution for a wide range of remote listening applications. I have found that meticulous attention to detail in all aspects of the system, from microphone selection to encryption protocols, is essential for achieving optimal performance and ensuring the integrity of the transmitted audio.

3.2 Networked Audio Systems

Let's delve into the world of networked audio systems. We're talking about getting sound from one place to another over a network, typically a digital network like the internet or a local area network (LAN). This opens up a huge range of possibilities, from simple home audio streaming to complex professional audio setups for broadcasting, recording studios, and live sound reinforcement. It's more than just plugging in a cable; it's about understanding the technologies and protocols that make it all work seamlessly.

One of the first hurdles you'll encounter is understanding the different protocols used to transmit audio over a network. While the underlying network might use standard TCP/IP, the audio itself needs to be packaged and delivered in a way that minimizes latency (delay) and ensures reliability. Several popular protocols have emerged to address these needs. Dante, developed by Audinate, is a widely adopted protocol in professional audio. It utilizes standard IP networks to transmit high-quality, low-latency audio. AVB (Audio Video Bridging) is another protocol that aims to provide guaranteed bandwidth and low latency for audio and video streams. AES67 is an open standard designed to achieve interoperability between different networked audio systems, allowing devices from different manufacturers to communicate with each other. Ravenna is yet another option, known for its high performance and flexibility. Each protocol has its own strengths and weaknesses, and the best choice for a particular application depends on factors such as the required audio quality, network infrastructure, and budget. Understanding these protocols is crucial for designing and implementing a networked audio system that meets your specific needs.

Then comes the selection of hardware for your networked audio system. This goes beyond just your microphones and speakers; it includes the devices that connect them to the network and manage the audio streams. Audio interfaces with network connectivity are essential. These devices convert analog audio signals from microphones or instruments into digital signals that can be transmitted over the network, and vice versa for sending audio to speakers or headphones. Network switches play a vital role in managing the flow of audio data across the network. For professional audio applications, managed switches are often preferred because they allow you to prioritize audio traffic and ensure that it receives sufficient bandwidth. Digital audio workstations (DAWs) can also be integrated into a networked audio system, allowing you to record, edit, and mix audio from multiple sources. Furthermore, dedicated audio processors with network connectivity can provide real-time signal processing capabilities, such as EQ, compression, and reverb, all controlled remotely over the network. Selecting the right hardware is a critical step in building a robust and reliable networked audio system.

Now, let's talk about the importance of network infrastructure. A robust and reliable network is the foundation of any successful networked audio system. Wireless networks, while convenient, can be prone to dropouts and latency issues, especially with multiple devices streaming audio simultaneously. A wired Ethernet connection is generally preferred for critical audio applications. Network switches should be of high quality and capable of handling the bandwidth requirements of the audio streams. Consider using Gigabit Ethernet switches to provide ample bandwidth for multiple audio channels. Quality of Service (QoS) settings on the network switch can be configured to prioritize audio traffic over other network traffic, such as web browsing or email, ensuring that audio streams receive preferential treatment. It's also crucial to properly configure the network to avoid IP address conflicts and other network-related issues that can disrupt audio streams. Paying attention to the network infrastructure is a fundamental requirement for a successful networked audio implementation.

I cannot stress enough the importance of latency. It's the enemy of networked audio, especially in live performance situations. Latency refers to the delay between when a sound is produced and when it is heard. Even a small amount of latency can be distracting and make it difficult for musicians to perform in sync. Several factors contribute to latency in networked audio systems, including the network infrastructure, the audio encoding and decoding processes, and the processing power of the devices involved. Minimizing latency requires careful planning and optimization. Choosing low-latency audio protocols, using high-performance network hardware, and optimizing the network configuration are all essential steps. Additionally, minimizing the number of hops that audio data must travel across the network can also help reduce latency. For live performance applications, latency should ideally be kept below 10 milliseconds to ensure a natural and responsive feel for the performers.

Synchronization, or the maintenance of a precise time alignment between different audio devices in the network, is another critical aspect to consider. Without proper synchronization, audio signals from different sources can drift out of sync, resulting in phasing issues and a loss of clarity. Networked audio systems typically rely on a master clock to synchronize all devices. The master clock provides a reference timing signal that all devices can use to align their audio samples. Precision Time Protocol (PTP) is a commonly used protocol for synchronizing networked audio devices. PTP allows for very accurate synchronization, typically within a few microseconds. Choosing a high-quality master clock and ensuring that all devices are properly synchronized to the master clock are essential for maintaining audio quality and clarity in a networked audio system.

Security should always be a priority. Networked audio systems are vulnerable to the same security threats as any other networked device, and it's important to take steps to protect them from unauthorized access. Use strong passwords for all devices and regularly update the firmware to patch security vulnerabilities. Firewalls can be used to restrict access to the audio network and prevent unauthorized devices from connecting. VLANs (Virtual LANs) can be used to segment the audio network from other network traffic, further isolating it from potential security threats. Consider implementing encryption to protect audio streams from eavesdropping. By taking these security measures, you can help ensure the privacy and integrity of your networked audio system.

To conclude, always conduct thorough testing! Before deploying a networked audio system, it's essential to test it thoroughly to ensure that it meets your requirements. Test the audio quality, latency, and reliability of the system under various conditions. Simulate real-world scenarios, such as multiple devices streaming audio simultaneously, to ensure that the network can handle the load. Monitor the network performance to identify any bottlenecks or potential issues. By testing the system thoroughly, you can identify and resolve any problems before they become major issues. Remember, a well-planned and tested networked audio system can provide a flexible, scalable, and reliable solution for a wide range of audio applications. So, go ahead, explore this exciting world, and unlock the power of networked audio.

4 Chapter 4: Perception and Psychoacoustics

Our ears are remarkably complex instruments, translating pressure waves into meaningful information about our surroundings. This chapter explores how our brains actively interpret these signals, a field known as psychoacoustics. It's not just about what sounds are present, but how we perceive them that matters in remote listening contexts. Understanding these perceptual nuances is crucial for designing systems that effectively convey information, particularly in challenging environments. This includes everything from designing better hearing aids to creating more effective surveillance systems.

The auditory system is not a passive receiver. From the moment sound enters the ear canal, a series of transformations occurs. The outer ear funnels sound towards the tympanic membrane (eardrum), which vibrates in response. These vibrations are amplified and transmitted through the middle ear via three tiny bones: the malleus, incus, and stapes. The stapes, connected to the oval window of the cochlea, sets the fluid within the cochlea in motion. This fluid movement stimulates hair cells, the sensory receptors within the cochlea, which are tuned to different frequencies. High frequencies are processed near the base of the cochlea, while low frequencies are processed at the apex. The hair cells convert this mechanical energy into electrical signals that travel along the auditory nerve to the brainstem, where initial processing begins. This initial processing involves comparing the signals from both ears to determine the direction and distance of the sound source. From the brainstem, the information travels to the auditory cortex in the temporal lobe, where higher-level processing occurs, allowing us to recognize and interpret sounds. This complex journey is not a simple transcription; it is an active process of filtering, amplification, and interpretation. For remote listening applications, understanding the limitations of this process is critical. For example, if the transmitted signal is distorted or incomplete, the brain must work harder to fill in the gaps, potentially leading to fatigue or misinterpretation.

One of the most fascinating aspects of auditory perception is our ability to focus on a single sound source amidst a cacophony of noise, commonly known as the "cocktail party effect." Imagine yourself at a crowded party, surrounded by numerous conversations, music, and clinking glasses. Yet, you can still focus on the person speaking directly to you, filtering out the irrelevant background noise. This remarkable ability relies on several factors. Spatial cues, such as the location of the sound source, play a crucial role. Our brains use the slight differences in timing and intensity of the sound arriving at each ear to pinpoint the location of the speaker. Additionally, our attention and expectations influence what we hear. If we are actively trying to listen to a particular conversation, we are more likely to filter out distracting sounds. The characteristics of the sound itself, such as its pitch, timbre, and rhythm, also contribute to our ability to segregate sound sources. For instance, a familiar voice is easier to distinguish from background noise than an unfamiliar one. The cocktail party effect demonstrates the brain's remarkable ability to selectively process auditory information. However, it also highlights the limitations of our auditory attention. In challenging listening environments, such as those encountered in remote surveillance or wildlife monitoring, even the most sophisticated algorithms may struggle to replicate the brain's ability to extract relevant information from noisy signals.

Masking is another crucial concept in psychoacoustics. It describes the phenomenon where one sound makes it difficult or impossible to hear another sound. The masking threshold is the minimum intensity at which a sound can be heard in the presence of a masking sound. Several factors influence masking thresholds, including the frequency, intensity, and duration of both the signal and the masker. Sounds with frequencies close to each other are more likely to mask each other than sounds with widely different frequencies. Louder sounds are more likely to mask quieter sounds. The temporal relationship between the signal and the masker also plays a role. For example, a masker that occurs immediately before the signal is more effective than one that occurs after the signal. Masking has significant implications for remote listening applications. In noisy environments, faint sounds of interest may be masked by louder background noise, making them difficult or impossible to detect. Understanding masking thresholds can help us design systems that minimize the effects of masking. This might involve using filtering techniques to remove the masking noise or amplifying the faint signals to make them more audible.

Furthermore, our perception of loudness is not linear. A sound that is twice as intense does not necessarily sound twice as loud. The relationship between intensity and perceived loudness is logarithmic. This is described by the Stevens' power law, which states that the perceived magnitude of a sensation is proportional to the physical intensity raised to a power. The exponent in Stevens' power law varies depending on the sensory modality. For loudness, the exponent is typically around 0.3. This means that a tenfold increase in intensity is only perceived as a doubling of loudness. Understanding this non-linear relationship is important for designing audio systems that accurately reproduce the dynamic range of sounds. It also helps us understand why some sounds are perceived as annoyingly loud, even if their intensity is not exceptionally high.

In addition to loudness, our perception of pitch is also complex. Pitch is primarily determined by the frequency of a sound, but other factors, such as intensity and timbre, can also influence our perception of pitch. The "missing fundamental" phenomenon demonstrates this complexity. When we hear a complex tone with a fundamental frequency and a series of harmonics, we perceive the pitch of the fundamental frequency, even if the fundamental frequency is not actually present in the sound. Our brains infer the fundamental frequency based on the pattern of harmonics. This phenomenon suggests that pitch perception is not simply a matter of detecting the lowest frequency component of a sound. It involves a more sophisticated process of pattern recognition. For remote listening applications, understanding pitch perception is important for accurately identifying and classifying sounds. For example, the pitch of a bird song can be used to identify the species of bird. The pitch of a human voice can be used to identify the speaker.

Finally, our auditory perception is highly adaptable. We can learn to distinguish subtle differences in sounds that were previously indistinguishable. This is known as perceptual learning. For example, experienced musicians can distinguish between extremely small differences in pitch. Trained listeners can identify subtle changes in the timbre of a sound. Perceptual learning demonstrates the plasticity of the auditory system. With practice and experience, we can improve our ability to perceive and interpret sounds. This has important implications for training people to use remote listening systems. With proper training, operators can learn to effectively filter out noise and extract relevant information from complex auditory signals.

4.1 Auditory Processing

The journey from sound wave to conscious perception is a complex and fascinating one. Our ears don't simply act as microphones; they are intricate biological instruments that perform a remarkable array of signal processing tasks before ever sending information to the brain. Understanding this processing is crucial for anyone working with remote listening technologies, because it dictates what information is actually available to the listener and how best to present it. The initial stage is the reception of sound waves by the outer ear, the pinna acting as a sound collector, funneling sound towards the ear canal. The unique shape of the pinna also plays a role in sound localization, especially in the vertical plane, as it slightly alters the frequency content of sounds depending on their elevation.

Next, the sound waves travel down the ear canal and cause the tympanic membrane, or eardrum, to vibrate. These vibrations are then transmitted through the middle ear by three tiny bones: the malleus, incus, and stapes, collectively known as the ossicles. This ossicular chain acts as an impedance matching device, amplifying the pressure of the sound waves to efficiently transfer energy from the air-filled middle ear to the fluid-filled inner ear. The stapes, the smallest bone in the human body, is connected to the oval window, an opening in the cochlea, the snail-shaped structure that houses the sensory receptors for hearing. The middle ear muscles, the stapedius and tensor tympani, also play a protective role, contracting in response to loud sounds to reduce the amount of energy transmitted to the inner ear. This is the acoustic reflex, which helps to protect the delicate structures of the cochlea from damage caused by excessive noise.

Within the cochlea, the mechanical vibrations are converted into electrical signals. This transduction process occurs in the organ of Corti, which contains the hair cells, the sensory receptors for hearing. As the stapes vibrates against the oval window, it creates pressure waves in the fluid-filled cochlea. These pressure waves cause the basilar membrane, a structure that runs along the length of the cochlea, to vibrate. The basilar membrane is tonotopically organized, meaning that different locations along its length respond maximally to different frequencies. High-frequency sounds cause the basilar membrane to vibrate most strongly near the base of the cochlea, while low-frequency sounds cause it to vibrate most strongly near the apex. This tonotopic organization is maintained throughout the auditory pathway, allowing the brain to distinguish between different frequencies. The hair cells are positioned on the basilar membrane, and as the membrane vibrates, the stereocilia, tiny hair-like projections on the tops of the hair cells, are deflected. This deflection opens mechanically gated ion channels, allowing ions to flow into the hair cells and causing them to depolarize. This depolarization triggers the release of neurotransmitters, which stimulate the auditory nerve fibers. There are two types of hair cells: inner hair cells and outer hair cells. The inner hair cells are primarily responsible for transmitting auditory information to the brain, while the outer hair cells play a role in amplifying and refining the vibrations of the basilar membrane.

Now that the auditory nerve fibers are activated, the electrical signals are transmitted along the auditory pathway to the brain. This pathway is complex and involves several relay stations, including the cochlear nucleus, superior olivary complex, inferior colliculus, and medial geniculate body. At each relay station, the auditory information is processed and refined. For example, the superior olivary complex is involved in sound localization, while the inferior colliculus integrates auditory information with information from other senses. Finally, the auditory information reaches the auditory cortex, located in the temporal lobe of the brain. The auditory cortex is responsible for the conscious perception of sound. Within the auditory cortex, different areas are specialized for processing different aspects of sound, such as pitch, loudness, and timbre. The processing isn't simply a hierarchical ascent; descending pathways also exist, allowing the brain to modulate the activity of the auditory system at lower levels.

The ear is not a perfect receiver; various psychoacoustic phenomena influence our perception of sound. Masking, for instance, occurs when one sound obscures another, making it difficult or impossible to hear. This is particularly relevant in remote listening scenarios where background noise may be present. The louder the masking sound, and the closer it is in frequency to the target sound, the more effective the masking will be. Another phenomenon is loudness perception, which is not a linear function of sound intensity. Our ears are more sensitive to some frequencies than others, and the perceived loudness of a sound depends on its frequency content. The equal-loudness contours, also known as Fletcher-Munson curves, illustrate this frequency-dependent sensitivity. Temporal masking also comes into play. A loud sound can mask quieter sounds that occur immediately before or after it. The pre-masking effect is usually shorter than the post-masking effect.

Consider the implications for remote monitoring of a patient's breathing. The subtle sounds of respiration might be easily masked by background noise in the hospital room, like the hum of equipment or conversations. Understanding masking thresholds can help engineers design systems that minimize noise interference and selectively amplify the respiratory sounds. Knowing these principles, we can use filters to reduce background sound and better isolate important sounds for the doctor to hear, leading to more accurate diagnoses. In a similar way, understanding the cocktail party effect allows us to build remote listening systems that help people focus on specific sounds, even in noisy places, which is

especially helpful for people with hearing problems.
Ultimately, grasping the intricacies of auditory processing provides a crucial lens through which to view the design and implementation of remote listening technologies. It allows for the creation of systems that not only capture and transmit sound effectively but also present it in a way that optimizes human perception and understanding. This knowledge is more than academic; it's the key to building tools that genuinely enhance our ability to connect with and understand the world around us, even from a distance. Therefore, a strong grounding in auditory science is indispensable for anyone aiming to contribute meaningfully to the field of remote listening.

4.2 The Cocktail Party Effect

Imagine yourself at a crowded gathering, a boisterous party buzzing with countless conversations, clinking glasses, and background music. It seems impossible to focus on any single voice amid the cacophony. Yet, you manage to tune into the person directly in front of you, filtering out the surrounding noise and understanding their words with relative ease. This remarkable feat of auditory selection is known as the "cocktail party effect," and it is far more complex than simple selective hearing. It represents a sophisticated interplay of sensory processing, cognitive attention, and prior knowledge, all working in concert to make sense of a complex auditory environment. In essence, it allows us to listen selectively, prioritizing specific sound sources while suppressing others, a crucial ability for communication and navigation in a world filled with competing sounds. This is particularly relevant in the context of remote listening, where the clarity and intelligibility of the target sound can be compromised by various background noises and environmental factors.
The cocktail party effect isn't just about turning up the volume on what you want to hear; it's a multi-layered process that begins with basic auditory processing. Our ears, equipped with specialized cells, detect the frequencies and amplitudes of incoming sound waves, converting them into neural signals that travel to the brainstem. Here, preliminary analysis begins, separating sounds based on basic acoustic features like pitch,

loudness, and location. This initial segregation is critical, as it provides the raw material for subsequent attentional mechanisms. For instance, if the person you're talking to has a distinctive voice, the brainstem might prioritize that frequency range, giving it preferential treatment in the next stages of processing. This automatic pre-attentive processing helps filter out sounds that are clearly irrelevant, reducing the cognitive load on higher brain regions. It's a constant process, working even when you're not consciously aware of it, laying the groundwork for focused listening. The brainstem acts as a first line of defense, swiftly categorizing and routing auditory information.

Once the sound signals reach the auditory cortex, a more detailed analysis takes place. The brain begins to construct a "soundscape," a comprehensive representation of all the auditory elements present in the environment. This soundscape includes not only the sounds themselves, but also their spatial relationships, temporal patterns, and contextual associations. This is where the cognitive magic really starts to happen. Your brain uses a combination of "bottom-up" and "top-down" processing to make sense of the sounds. Bottom-up processing refers to the analysis of the acoustic features of the sound itself – its pitch, timbre, and rhythm. Top-down processing, on the other hand, relies on your prior knowledge, expectations, and goals. For example, if you know you're at a party, your brain will automatically activate a schema for "party sounds," which might include music, laughter, and conversations. This schema helps you interpret the incoming sounds and distinguish between relevant and irrelevant information. In the context of the cocktail party effect, top-down processing allows you to anticipate what the person you're talking to is likely to say, filling in gaps in their speech and understanding their meaning even when their words are partially obscured by noise. Essentially, the brain doesn't just passively receive sounds; it actively constructs a meaningful representation of the auditory world.

Attention plays a crucial role in the cocktail party effect. It acts like a spotlight, selectively amplifying the neural signals associated with the target sound source while suppressing the signals from competing sounds. This attentional modulation can occur at multiple levels of the auditory system, from the brainstem to the auditory cortex. There are two main types of attention involved: selective attention and divided attention. Selective attention refers to the ability to focus on a single sound source while ignoring others. Divided attention, on the other hand, refers to the ability to attend to multiple sound sources simultaneously. In the cocktail party effect, you primarily rely on selective attention to focus on the person you're talking to. However, divided attention also comes into play, as you might occasionally monitor other conversations or sounds in the environment to maintain situational awareness. Furthermore, attention is not a static process; it fluctuates dynamically over time, shifting focus as needed to respond to changes in the auditory environment. This flexibility is essential for navigating complex soundscapes and extracting the information you need.

Beyond attention, memory and learning are also critical components of the cocktail party effect. Your brain uses past experiences and learned associations to interpret the sounds you hear. For example, if you've had many conversations with a particular person, you'll be better able to understand their speech even in noisy environments, because you've learned their unique vocal characteristics and speaking style. Similarly, if you're familiar with a particular accent or dialect, you'll find it easier to understand speech spoken in that accent, even if it's partially masked by noise. This reliance on memory and learning highlights the importance of experience in shaping our auditory perception. The more experience you have with a particular sound environment, the better you'll become at filtering out irrelevant noise and focusing on the information you need. This also applies to recognizing and distinguishing between different types of background noise; the more familiar you are with a particular noise profile, the easier it will be to ignore it.

Applying this understanding to remote listening technologies is crucial. When developing remote listening systems, it is important to consider the challenges posed by the cocktail party effect. This means designing systems that minimize background noise, enhance the clarity of the target sound, and present information in a way that is easily understood by the listener. For instance, noise reduction algorithms can be used to suppress unwanted sounds, while directional microphones can be used to focus on the target sound source. Furthermore, the user interface of the remote listening system should be designed to minimize cognitive load and facilitate selective attention. This might involve providing visual cues to help the listener focus on the target sound, or allowing the listener to adjust the volume and equalization of different sound sources. By taking into account the principles of the cocktail party effect, we can develop remote listening systems that are more effective, user-friendly, and less prone to errors caused by auditory distractions. Ultimately, the goal is to create a seamless and intuitive listening experience, enabling users to extract the information they need even in challenging acoustic environments.

4.3 Masking Thresholds

Imagine you're in a crowded marketplace. A cacophony of sounds assaults your ears – vendors hawking their wares, children laughing, music playing, animals bleating, and the general hubbub of human activity. Yet, somehow, you can still focus on the conversation you're having with a friend, filtering out the surrounding noise to understand their words. This remarkable feat is partly due to a phenomenon called "masking," and the level at which one sound obscures another is known as the masking threshold. Understanding these thresholds is paramount in the field of remote listening, where clarity and accuracy are crucial.

In essence, masking occurs when one sound makes it difficult or impossible to hear another sound. This isn't simply about loudness; it's about the interplay of frequencies, intensities, and timing. A loud truck rumbling past might drown out a whisper, but a high-pitched whine might be easily heard despite the roar. This is because the masking effect is frequency-dependent. A tone will most effectively mask other tones close to it in frequency. This critical band theory suggests that our auditory system analyzes sounds within specific frequency ranges, and interference within these bands is most likely to cause masking. This means that if you're trying to remotely listen to a specific frequency range, it's crucial to understand the ambient noise profile and identify potential masking frequencies. The closer the frequencies of the masker and the target, the stronger the masking effect. It's why a high-pitched alarm can cut through low-frequency background noise more effectively than a rumble of similar loudness.

Several factors influence the masking threshold. The intensity of the masking sound is, of course, a primary determinant. Louder sounds are generally more effective at masking quieter sounds. However, the frequency content of the masking sound is also significant. Broadband noise, containing a wide range of frequencies, is generally more effective at masking than a single tone. Temporal masking, another crucial aspect, refers to how sounds can mask other sounds that occur before or after them. Forward masking occurs when a sound masks another sound that immediately follows it, while backward masking happens when a sound masks a sound that immediately precedes it. The duration and interval between the sounds play significant roles in temporal masking. For example, a sudden, loud burst of noise can mask quieter sounds for a brief period afterward, making it difficult to discern subtle details in the immediate aftermath. Understanding temporal masking is particularly important in dynamic acoustic environments where sounds are constantly changing.

Consider a practical example in the context of wildlife monitoring. Imagine you're using remote microphones to listen for the calls of a specific bird species in a forest. The rustling of leaves in the wind, the buzzing of insects, and the distant rumble of traffic all contribute to the overall noise floor. If the bird's calls are relatively quiet and fall within a frequency range similar to the background noise, they may be masked, rendering them undetectable. To overcome this, you might employ various signal processing techniques. Frequency-selective filtering can attenuate the noise in the relevant frequency bands, effectively lowering the masking threshold. Alternatively, techniques like spectral subtraction can estimate the noise profile and subtract it from the overall signal, revealing the underlying bird calls. Directional microphones can also be used to focus on the area where the bird is most likely to be, reducing the amount of background noise picked up.

Another application lies in medical diagnostics, specifically remote auscultation – listening to internal body sounds using a stethoscope remotely. Heart sounds, lung sounds, and bowel sounds are often subtle and can be easily masked by ambient noise in the environment. In a noisy emergency room, for example, it can be difficult to discern faint heart murmurs. To improve the clarity of these sounds, noise-canceling headphones, advanced filtering techniques, and even acoustic isolation chambers can be employed. Furthermore, understanding the frequency characteristics of different body sounds allows for targeted filtering to reduce masking from noise sources. For instance, heart sounds typically fall in the lower frequency range, while lung sounds extend to higher frequencies.

Beyond the specific techniques, a fundamental step is always to characterize the acoustic environment. Before deploying any remote listening system, it's crucial to perform a thorough analysis of the typical noise levels and frequency characteristics at the intended location. This involves recording ambient noise at different times of day and under different environmental conditions. This data can be used to create a noise profile, which informs the selection of appropriate microphones, filters, and signal processing algorithms. Moreover, it allows for setting appropriate recording thresholds and gain levels to maximize the signal-to-noise ratio and minimize the risk of masking. If, for example, you are monitoring a busy street for specific sound events, understanding the peak hours of traffic noise will allow you to adjust the system to effectively capture these events.

Effective strategies for minimizing the impact of masking involve a combination of acoustic design, signal processing, and attentive monitoring. Consider the acoustic properties of the environment where the listening device is deployed. Minimizing reflective surfaces and using sound-absorbing materials can reduce reverberation and ambient noise levels. Deploying microphones in locations with less ambient noise, such as shielded enclosures or quiet zones, can significantly improve the signal-to-noise ratio. Then, signal processing techniques like adaptive filtering can dynamically adjust to changing noise conditions, effectively removing unwanted noise while preserving the target sounds. Finally, continuous monitoring of the audio signal can provide real-time feedback on the effectiveness of the system and allow for adjustments to be made as needed. Analyzing spectrograms, for instance, can visually highlight frequency ranges where masking is occurring, allowing for targeted intervention.

Masking thresholds are not static; they are influenced by individual hearing abilities. People with hearing loss, for instance, often have elevated masking thresholds, making it more difficult for them to hear sounds in noisy environments. This has significant implications for the design of assistive listening devices. These devices must be carefully tuned to compensate for individual hearing impairments and minimize the effects of masking. Furthermore, the age of the listener can also affect masking thresholds, with older individuals typically experiencing greater difficulty in discerning sounds in the presence of noise. This necessitates personalized approaches to audio processing in remote listening applications, tailoring the amplification and filtering to the specific needs of the user. In the world of remote listening, where subtle sounds can hold vital information, mastering the principles of masking thresholds is paramount. It's about more than just turning up the volume; it's about understanding the complex interplay of sound waves, the human ear, and the surrounding environment. By combining a solid understanding of psychoacoustics with practical signal processing techniques, we can overcome the challenges of masking and unlock the full potential of remote auditory perception. In doing so, we can build systems that not only hear, but also truly understand.

5 Chapter 5: Environmental Influences

The real world isn't a laboratory, and when attempting to listen remotely, the pristine, clear signal you might expect often dissolves into a cacophony of background noise and environmental distortions. Understanding and mitigating these influences is paramount to the success of any remote listening endeavor. We must first acknowledge the pervasiveness of noise pollution. It's not merely the obvious clamor of urban environments; it's the hum of electrical equipment, the rustle of wind through trees, the subtle vibrations of machinery. Each of these contributes to a complex soundscape that can obscure or distort the target sounds.

Noise pollution analysis begins with identifying the dominant noise sources in a particular environment. This is not as simple as it sounds. Sophisticated tools and techniques are required to decompose the acoustic environment into its constituent parts. Sound level meters, spectral analyzers, and directional microphones are indispensable for quantifying the intensity, frequency, and directionality of different noise sources. Beyond simple measurement, advanced signal processing techniques, such as independent component analysis (ICA), can be used to separate overlapping sound sources, even when they share similar frequency characteristics. Consider a scenario where you're attempting to monitor the health of a mechanical system remotely. The engine's vibrations, the hiss of hydraulics, and the ambient hum of the factory floor all contribute to a complex acoustic mix. ICA can help you isolate the unique sound signature of a failing bearing, which might otherwise be masked by the overall noise. But it requires a deep understanding of acoustic signal processing and proper selection of its parameters.

Acoustic modeling tools take this analysis a step further, enabling us to predict how sound will propagate through a given environment. These tools utilize sophisticated algorithms to simulate the complex interactions of sound waves with surfaces, obstacles, and atmospheric conditions. Ray tracing, finite element analysis (FEA), and boundary element methods (BEM) are commonly employed to create detailed acoustic models. Ray tracing, for example, simulates sound propagation by tracing the paths of numerous "rays" from a sound source, accounting for reflections, diffractions, and absorption along each path. FEA and BEM, on the other hand, solve the wave equation numerically, providing a more accurate, but computationally intensive, representation of the sound field. Imagine you're trying to monitor wildlife in a dense forest. The trees, the terrain, and even the atmospheric conditions will all affect how sound travels. An acoustic model can help you optimize the placement of microphones to maximize the likelihood of detecting the target sounds, such as the calls of a particular bird species. Furthermore, modeling can show the effect of temperature gradients on sound propagation, making it possible to adjust the setup according to expected weather conditions.

But modeling isn't a perfect science. It relies on accurate input data, including the acoustic properties of the materials in the environment. These properties, such as the absorption coefficient and scattering coefficient, describe how a material interacts with sound waves. Obtaining accurate measurements of these properties can be challenging, often requiring specialized equipment and expertise. Moreover, acoustic models are only as good as the assumptions they make. Real-world environments are often complex and unpredictable, with factors such as wind, humidity, and temperature variations that can significantly affect sound propagation. This means that acoustic models should be viewed as a guide, rather than a definitive prediction. Validating the model with real-world measurements is always essential.

To effectively combat environmental noise, we must employ a range of strategies. Noise reduction techniques can be broadly categorized into passive and active methods. Passive methods involve physically blocking or absorbing sound, using materials with high absorption coefficients, such as acoustic foam, heavy curtains, or strategically placed barriers. These are appropriate if the main sound source to block is well-defined. Active noise cancellation (ANC) technologies, on the other hand, use electronic means to generate sound waves that are 180 degrees out of phase with the unwanted noise, effectively canceling it out. ANC is particularly effective for reducing low-frequency noise, such as the hum of machinery or the drone of aircraft. However, it requires precise control of the phase and amplitude of the canceling signal, and it can be less effective in complex acoustic environments with multiple noise sources. For instance, in a remote monitoring application near a busy road, a combination of a physical barrier (passive) and ANC (active) might be necessary to achieve optimal noise reduction.

Signal processing algorithms play a crucial role in enhancing the clarity of remote audio. Adaptive filtering techniques can be used to estimate and remove noise components from the received signal. These filters "learn" the characteristics of the noise and adjust their parameters accordingly, allowing them to adapt to changing noise conditions. Spectral subtraction is another common technique, which involves estimating the power spectrum of the noise and subtracting it from the power spectrum of the received signal. However, spectral subtraction can sometimes introduce artifacts, such as "musical noise," which can be distracting to the listener. More advanced techniques, such as Wiener filtering and Kalman filtering, offer improved noise reduction performance, but they require more computational resources. Imagine you're monitoring a conversation in a public space. The audio signal is likely to be contaminated by background noise, such as traffic noise, conversations, and music. Adaptive filtering can help you isolate the target conversation, making it easier to understand what is being said. But you need to select the right settings, or risk distorting the speech itself.

Ultimately, the success of remote listening depends on a holistic approach that considers all aspects of the acoustic environment. This includes careful site selection, strategic microphone placement, effective noise reduction techniques, and sophisticated signal processing algorithms. It also requires a deep understanding of the limitations of each technique and the trade-offs involved. By combining these elements, we can overcome the challenges posed by environmental influences and achieve clear, reliable remote audio monitoring. It's not just about having the best technology; it's about understanding how that technology interacts with the world around it, and adapting accordingly.

5.1 Noise Pollution Analysis

Noise pollution presents a significant challenge to the effectiveness of any remote listening system. Its insidious presence permeates virtually every environment, corrupting audio signals and obscuring the very sounds we seek to capture. Understanding its characteristics and developing strategies to mitigate its impact are crucial for achieving reliable remote listening.

We can begin by acknowledging that noise is subjective; what constitutes noise for one application might be perfectly acceptable, or even desirable, in another. However, for our purposes, noise is any unwanted sound that interferes with the clarity and intelligibility of the target signal. Noise pollution, therefore, encompasses a wide spectrum of sounds, from the low-frequency rumble of traffic and machinery to the high-pitched whine of electronics and the cacophony of human conversation. The sources can be classified into categories: transportation noise (vehicles, aircraft, trains), industrial noise (machinery, construction), domestic noise (appliances, music, neighbors), and environmental noise (wind, rain, animal sounds). Each category presents unique challenges and requires tailored approaches for analysis and mitigation.

Quantifying noise pollution is the next essential step. This involves measuring sound levels using specialized instruments like sound level meters. These meters provide a snapshot of the instantaneous sound pressure level, typically expressed in decibels (dB). However, a single measurement is rarely sufficient to characterize the overall noise environment. Instead, it's necessary to collect data over time, capturing the variations in noise levels throughout the day and night. The equivalent continuous sound level (Leq) is a commonly used metric that represents the average sound energy over a specific period. Another useful metric is the Ldn (day-night average sound level), which takes into account the increased sensitivity to noise during nighttime hours by applying a penalty to noise levels occurring between 10 PM and 7 AM. By collecting and analyzing these data, we can gain a clear understanding of the dominant noise sources, their temporal patterns, and their overall contribution to the acoustic environment.

The frequency content of noise is as important as its amplitude. A spectral analysis, typically performed using a spectrum analyzer, reveals the distribution of sound energy across different frequency bands. This information is invaluable for identifying specific noise sources and designing effective filtering strategies. For example, if a remote listening system is intended to capture speech, we can focus on attenuating noise in frequency bands outside the typical speech range (e.g., below 300 Hz or above 4 kHz). Understanding the spectral characteristics of noise also helps in selecting appropriate microphone types and placement techniques. Microphones with specific directional patterns can be used to minimize the pickup of noise from certain directions. Furthermore, strategic placement of the microphone can take advantage of acoustic shielding or distance attenuation to reduce the impact of noise sources.

Acoustic modeling tools play a vital role in predicting and analyzing the propagation of noise in complex environments. These tools use computer simulations to estimate sound levels at different locations based on factors such as the geometry of the environment, the location and characteristics of noise sources, and the properties of the materials present. Acoustic modeling can be used to optimize the placement of microphones, evaluate the effectiveness of noise control measures, and assess the potential impact of noise on the performance of remote listening systems. These models let us virtually test different configurations, noise barriers, sound-absorbing materials, to minimize the disturbances.

Noise reduction is essential for a proper use. Techniques include passive and active methods. Passive noise control involves the use of physical barriers or sound-absorbing materials to reduce the transmission of noise. Examples include soundproof enclosures, noise barriers along roadways, and acoustic panels in rooms. Active noise control (ANC) uses electronic systems to generate sound waves that are out of phase with the noise, effectively canceling it out. ANC systems are often used in headphones and other personal audio devices to reduce background noise. For remote listening applications, ANC can be implemented near the microphone to improve the signal-to-noise ratio.

Signal processing techniques are also crucial. Filtering is a fundamental technique for removing unwanted noise from audio signals. Various types of filters can be used, depending on the characteristics of the noise. Low-pass filters attenuate high-frequency noise, while high-pass filters attenuate low-frequency noise. Band-pass filters allow a specific range of frequencies to pass through, while attenuating frequencies outside that range. Adaptive filtering techniques can dynamically adjust the filter characteristics based on the changing noise environment. These filters use algorithms to estimate the noise signal and subtract it from the desired signal, resulting in improved clarity.

Beyond the direct manipulation of audio signals, we should consider more holistic approaches to noise pollution. This includes careful site selection for remote listening devices, avoiding locations with high levels of noise pollution. Furthermore, educating the public about the impact of noise pollution can promote responsible behavior and reduce the overall noise levels in communities. By combining technical expertise with social awareness, we can create a quieter and more conducive environment for remote listening.

Finally, continuous monitoring and evaluation are essential for maintaining the effectiveness of noise mitigation strategies. Regular sound level measurements should be conducted to track changes in the noise environment and assess the performance of noise control measures. The feedback from the users of remote listening systems should be obtained to identify any remaining noise-related issues. The data collected from these evaluations should be used to refine and improve the noise mitigation strategies, ensuring that remote listening systems continue to provide clear and intelligible audio signals. Only then, noise pollution can be fully controlled.

5.2 Acoustic Modeling Tools

Acoustic modeling tools represent a crucial component in the success of any remote listening application, enabling us to predict and understand how sound behaves in a particular environment before even deploying any actual equipment. This proactive approach allows for the optimization of microphone placement, selection of appropriate equipment, and the implementation of signal processing techniques that effectively mitigate the effects of environmental noise and distortion. These tools are essential because the real world is far from a controlled laboratory setting; sounds bounce off surfaces, get absorbed by materials, and are distorted by atmospheric conditions.

The foundation of acoustic modeling relies on the principles of wave acoustics and numerical methods. One of the most common approaches is the Finite Element Method (FEM), which divides a complex space into smaller, simpler elements and solves acoustic equations, such as the Helmholtz equation, within each element. FEM is particularly adept at handling complex geometries and boundary conditions, making it suitable for modeling indoor environments with intricate architectural details. Another popular technique is the Boundary Element Method (BEM), which focuses on the surfaces of objects within the acoustic space. BEM is computationally efficient for exterior acoustics problems, like predicting noise propagation outdoors. Ray tracing, a geometric acoustics method, simulates sound propagation by tracing rays from a source, reflecting and diffracting them as they encounter surfaces. It's computationally less intensive than FEM or BEM, making it suitable for large-scale environments, though it is less accurate at low frequencies where wave effects are dominant. These fundamental methods can be used in any acoustic environment.

Specialized software packages, such as COMSOL Multiphysics, ANSYS, and specialized acoustic simulation tools like ODEON and EASE, provide user-friendly interfaces and advanced features for acoustic modeling. These packages allow users to import CAD models of the environment, assign acoustic properties to materials (absorption coefficients, scattering coefficients, etc.), define sound sources, and simulate sound propagation.

For example, you might use COMSOL to model the acoustic behavior of a concert hall, taking into account the shape of the walls, the materials used in the construction, and the placement of loudspeakers. The software can then predict the sound pressure levels and reverberation times at various locations within the hall, allowing you to optimize the acoustic design. It is important to consider the acoustic environment.

However, the accuracy of these models depends heavily on the quality of the input data. Precise measurements of material properties, such as absorption coefficients and scattering coefficients, are essential. These properties can be obtained from specialized databases or through laboratory measurements using impedance tubes or reverberation chambers. Similarly, accurate geometric models of the environment are crucial. While CAD models are ideal, they may not always be available. In such cases, laser scanning or photogrammetry can be used to create detailed 3D models of the space. Careful attention should be paid to the resolution of the model, as finer details can significantly impact the accuracy of the simulation, especially at higher frequencies. It is important to have quality data to run simulations.

Once the model is built and validated, it can be used to optimize various aspects of the remote listening system. Microphone placement is a critical factor affecting the quality of the captured audio. Acoustic modeling can help identify locations with optimal signal-to-noise ratio and minimal distortion. For instance, in a noisy industrial environment, the model can predict the sound pressure levels from the target source and the background noise at different locations, allowing you to strategically place microphones to maximize the clarity of the desired signal. Furthermore, modeling can inform the selection of appropriate microphones. Some microphones are more directional than others, which means they are more sensitive to sounds arriving from a specific direction while attenuating sounds from other directions. By simulating the sound field, you can determine the optimal microphone directivity pattern to minimize the pickup of unwanted noise.

Signal processing techniques can be tailored to the specific acoustic characteristics of the environment. For example, if the model reveals that certain frequencies are significantly attenuated due to absorption or interference, you can design equalization filters to compensate for these losses. Similarly, if reverberation is a major issue, you can implement dereverberation algorithms to remove the unwanted echoes and improve the clarity of the signal. Noise reduction techniques, such as adaptive filtering or spectral subtraction, can be used to suppress background noise based on the predicted noise characteristics. These techniques are important.

To make the most of acoustic modeling tools, it's essential to adopt a systematic approach. First, clearly define the objectives of the modeling exercise. What specific questions are you trying to answer? Are you trying to optimize microphone placement, select appropriate equipment, or design signal processing algorithms? Second, gather accurate input data, including geometric models and material properties. Don't underestimate the importance of this step, as the accuracy of the model depends heavily on the quality of the input data. Third, build the model using appropriate software and techniques. Start with a simplified model and gradually increase the complexity as needed. Fourth, validate the model by comparing its predictions with real-world measurements. This step is crucial for ensuring the reliability of the model. Finally, use the model to answer your questions and optimize the remote listening system. Remember that acoustic modeling is an iterative process; you may need to refine the model and repeat the simulations several times to achieve the desired results. It is important to be iterative.

Consider the practical challenges that may arise when using these tools. Computational resources can be a limiting factor, especially for complex models and large-scale environments. In such cases, you may need to simplify the model or use more powerful hardware. Data acquisition can also be challenging, particularly when measuring material properties in situ. Furthermore, the accuracy of the model is limited by the accuracy of the underlying acoustic equations and numerical methods. At low frequencies, wave effects can be complex and difficult to model accurately. Despite these challenges, acoustic modeling tools are invaluable for optimizing remote listening systems, enabling us to capture and analyze sound in a wide range of environments. Don't forget that there are a range of challenges that can occur.

Finally, continuous learning and staying updated with the latest advancements in acoustic modeling is essential for maximizing the benefits of these tools. New techniques and software packages are constantly being developed, offering improved accuracy, efficiency, and ease of use. By actively participating in the acoustic community, attending conferences, reading research papers, and experimenting with new tools, you can stay ahead of the curve and leverage the full potential of acoustic modeling in your remote listening applications. Remember that technology advances at a fast pace and that is important to be always up-to-date.

6 Chapter 6: Applications and Implications

Remote listening technology isn't just a futuristic concept anymore; it's weaving its way into the fabric of our everyday lives, albeit often behind the scenes. Think beyond the stereotypical spy movies and consider the sheer breadth of its applications, from safeguarding endangered species to revolutionizing healthcare. It's a landscape ripe with possibilities, but one we need to navigate carefully, ensuring its power is used responsibly. Consider, for instance, the burgeoning field of wildlife monitoring. Conservationists are increasingly relying on strategically placed microphones to eavesdrop on the natural world. These devices, often solar-powered and remotely accessible, can detect the subtle sounds of animal calls, footsteps, or even the rustling of leaves, providing invaluable insights into population dynamics, migration patterns, and the overall health of ecosystems. Imagine researchers in a lab thousands of miles away, monitoring the distress calls of elephants being poached or tracking the elusive movements of a snow leopard across a vast, mountainous terrain. The possibilities are truly transformative. The data captured is analyzed using advanced sound-recognition algorithms, some of which can identify individual animals based on their unique vocalizations, creating a kind of acoustic fingerprint. This allows researchers to track specific individuals over long periods, providing a detailed understanding of their behavior and social interactions, all without direct human intervention.

Moving from the wild to the clinic, the potential for remote listening in medical diagnostics is equally compelling. Telemedicine is already transforming access to healthcare, particularly in remote or underserved communities, and remote auscultation – the practice of listening to internal body sounds – is playing a significant role. Equipped with specialized digital stethoscopes and sophisticated software, doctors can listen to heart and lung sounds from afar, detecting abnormalities that might otherwise go unnoticed. This is particularly valuable for patients with chronic conditions who require regular monitoring, reducing the need for frequent and often expensive hospital visits. Furthermore, remote auscultation can be used to triage patients in emergency situations, allowing paramedics to transmit vital signs to specialists in real time, enabling faster and more informed decision-making. The development of miniaturized, wearable sensors is further expanding the possibilities, allowing for continuous, unobtrusive monitoring of vital signs, providing doctors with a comprehensive view of a patient's health over time.

Beyond wildlife and health, surveillance technologies represent a more complex and often controversial application. While surveillance can be used for legitimate purposes, such as crime prevention and security, it also raises serious concerns about privacy and civil liberties. Remote listening devices are increasingly being deployed in public spaces, such as streets, parks, and transportation hubs, to detect potential threats, monitor crowd behavior, and gather evidence in criminal investigations. However, the widespread use of such technology raises the specter of mass surveillance, where individuals are constantly being monitored and their privacy is compromised. It is crucial to establish clear legal frameworks and ethical guidelines to govern the use of surveillance technologies, ensuring that they are used responsibly and that individuals' rights are protected. Public transparency and accountability are essential to maintain public trust and prevent the abuse of power.

Assistive listening devices represent a powerful and positive application of remote listening, enhancing the lives of individuals with hearing impairments. These devices, ranging from simple personal amplifiers to sophisticated wireless systems, can amplify and clarify sounds, making it easier for people with hearing loss to communicate, participate in social activities, and enjoy everyday experiences. Furthermore, assistive listening technology is becoming increasingly integrated into smartphones and other mobile devices, providing users with personalized audio settings and real-time transcription capabilities. This not only improves their ability to hear and understand speech but also empowers them to take control of their listening environment, reducing the stigma associated with hearing loss and promoting greater independence. The development of advanced noise-cancellation algorithms and directional microphones is further enhancing the effectiveness of assistive listening devices, allowing users to focus on desired sounds while minimizing background noise.

Addressing security protocols is paramount given the sensitivity of data transmitted and collected through remote listening devices. Securing these systems is a multi-faceted challenge. Encryption plays a crucial role, scrambling the audio data during transmission to prevent unauthorized access. Strong authentication mechanisms are also essential, ensuring that only authorized users can access and control the devices. Regular security audits and vulnerability assessments are necessary to identify and address potential weaknesses in the system. Physical security measures, such as tamper-proof casings and secure mounting hardware, can also help prevent unauthorized access and modification of the devices. In addition to technical measures, robust policies and procedures are needed to govern the handling and storage of audio data, ensuring that it is protected from unauthorized disclosure or misuse. The implementation of strong data governance frameworks, coupled with ongoing employee training and awareness programs, can help minimize the risk of security breaches and data leaks. These protocols must adapt to ever-evolving cyber threats, requiring constant vigilance and investment in security measures.

The implications of remote listening technology extend far beyond these specific examples. As the technology continues to evolve and become more sophisticated, its potential applications will only expand. We must carefully consider the ethical, social, and legal implications of these advancements, ensuring that the technology is used in a way that benefits society as a whole. A collaborative, interdisciplinary approach, involving technologists, ethicists, policymakers, and the public, is essential to guide the responsible development and deployment of remote listening technology. Only through careful planning and thoughtful consideration can we harness its full potential while mitigating the risks. The future of remote listening is not predetermined; it is up to us to shape it.

6.1 Surveillance Technologies

The world of surveillance technologies is a complex landscape, constantly evolving with advancements in acoustics, signal processing, and communication. These tools, often cloaked in secrecy and shrouded in ethical dilemmas, leverage the principles we've discussed to extract information from sound, sometimes without our explicit knowledge or consent. Let's delve into the specific mechanisms and their implications, building on the foundations laid in previous chapters.

At its core, acoustic surveillance relies on the same principles as any other remote listening application – capturing, transmitting, and interpreting sound. However, the intent and deployment differ significantly. While wildlife monitoring might focus on capturing the subtle chirps of endangered birds, surveillance aims to capture human speech, environmental sounds indicating activity, or even mechanical noises from equipment. This requires highly sensitive microphones, strategically placed to maximize signal acquisition while minimizing detectability. Miniaturization is key here. Think of the tiny electret condenser microphones, often hidden within everyday objects or embedded in seemingly innocuous devices. These are designed for low power consumption and can be easily concealed, providing a stealthy means of capturing audio. However, the placement isn't random, it requires a deep understanding of acoustic principles, similar to what we explored in Chapter 5 on environmental influences. Placement takes into consideration the reflective surfaces, the natural noise, and common frequencies. The goal is to have the clearer recording as possible.

Beyond the microphones themselves, the transmission methods employed in surveillance technologies are also critical. Early surveillance relied heavily on wired connections, limiting their range and increasing the risk of detection. Today, wireless technology dominates. Bluetooth, Wi-Fi, and cellular networks offer convenient and often undetectable channels for transmitting audio data. The use of spread spectrum techniques, which distribute the signal across a wider frequency range, makes interception more difficult. Similarly, encryption plays a crucial role in protecting the confidentiality of the audio data. Advanced Encryption Standard (AES) and other robust encryption algorithms are employed to scramble the audio signal, rendering it unintelligible to unauthorized listeners. However, the implementation is the key, as even the strongest encryption can be broken with enough resources and time. This is where signal processing comes into play. Sophisticated algorithms can analyze the audio stream, identifying patterns and extracting relevant information, even in noisy environments. Voice activity detection (VAD) algorithms can identify periods of speech, allowing the system to focus on those segments and discard irrelevant background noise. Speaker recognition algorithms can attempt to identify individuals based on their voice characteristics, though these are not foolproof and are heavily influenced by the quality of the audio and the presence of background noise. These algorithms are also trained on specific data sets, and their accuracy can vary depending on the speaker's accent, age, and other factors.

Furthermore, the advancements in networked audio systems, as discussed in Chapter 3, have dramatically expanded the capabilities of surveillance. IP-based audio surveillance systems can transmit audio data over vast distances, allowing for centralized monitoring and analysis. Cloud storage provides virtually unlimited storage capacity for captured audio, enabling long-term archiving and analysis. This raises significant privacy concerns, as vast amounts of audio data can be collected and stored indefinitely, potentially being used for purposes beyond their original intent. Imagine a scenario where microphones are installed in public spaces under the guise of security measures, but the collected audio data is also used for marketing purposes or to track individual movements. The ethical implications are staggering.

Legal frameworks are struggling to keep pace with these technological advancements. Many countries have laws regulating wiretapping and electronic surveillance, but these laws often predate the widespread use of wireless technology and the Internet. The interpretation and enforcement of these laws can be complex and vary significantly across jurisdictions. A key distinction is often made between intercepting communications and accessing stored communications. Intercepting a phone call in real-time typically requires a warrant, while accessing a voicemail message stored on a server may be subject to different legal standards. However, the line between these two categories is becoming increasingly blurred as technology evolves. For instance, a system that records and analyzes audio from a smart speaker in a home blurs the line between interception and storage.

Counter-surveillance measures are also evolving. Techniques like white noise generators can mask conversations, making it difficult to capture intelligible audio. Acoustic jammers emit a high-intensity sound signal that overpowers any nearby microphones. However, these methods are not always effective and can be easily detected. A more sophisticated approach involves using acoustic metamaterials to create "cloaking devices" that deflect sound waves, making it difficult to detect the presence of a microphone. These technologies are still in their early stages of development, but they hold promise for protecting privacy in a world increasingly saturated with surveillance devices. Moreover, understanding the vulnerabilities inherent in wireless transmission and encryption protocols is crucial for mitigating the risk of eavesdropping. Regularly updating software and firmware, using strong passwords, and being cautious about connecting to public Wi-Fi networks can help protect against unauthorized access to audio data.

The future of surveillance technologies is likely to see even greater integration with sensor networks, cameras, and data analytics. Imagine a system that combines audio surveillance with facial recognition and behavioral analysis to identify potential threats in real-time. While such systems could potentially enhance security and prevent crime, they also raise profound concerns about mass surveillance and the erosion of privacy. The key to navigating this complex landscape is to promote transparency, accountability, and ethical considerations in the development and deployment of these technologies. We must ensure that the benefits of surveillance are not achieved at the expense of fundamental human rights.

6.2 Wildlife Monitoring

Imagine walking through a dense forest, the air thick with humidity, the ground soft beneath your boots. The symphony of nature surrounds you – the rustling of leaves, the chirping of insects, the distant calls of birds. But what if you could capture this soundscape remotely, without disturbing the delicate balance of the ecosystem? This is the essence of wildlife monitoring using remote listening technology.

One of the first steps in establishing a successful wildlife monitoring program is identifying your target species and their unique acoustic signatures. Each animal has its own distinct vocalizations, whether it's the hoot of an owl, the howl of a wolf, or the croak of a frog. By understanding these individual sound characteristics, you can effectively pinpoint and track their presence within a given area. This isn't always straightforward; sometimes, you'll need to filter out extraneous noise or learn to differentiate between similar sounds made by different species. Think of it as learning a new language – the language of the wild. Careful observation in the field, coupled with detailed recordings and spectral analysis, allows you to create a comprehensive library of acoustic identifiers.

Next, consider the selection and placement of your recording equipment. There's no one-size-fits-all solution; the optimal choice depends on the terrain, weather conditions, and the specific behaviors you're trying to capture. For example, directional microphones are excellent for isolating sounds from a specific location, such as a bird's nest, while omnidirectional microphones capture a broader range of sounds across a larger area. Similarly, weatherproof enclosures are essential in harsh environments to protect sensitive equipment from rain, snow, and extreme temperatures. The placement of these devices is equally important. Strategically position them near known animal habitats, migration routes, or breeding grounds to maximize the chances of capturing relevant data.

Once your equipment is in place, data management becomes paramount. The sheer volume of audio recordings can quickly become overwhelming, so it's crucial to establish a systematic approach to storage, organization, and analysis. This often involves utilizing specialized software that can automatically detect and classify animal vocalizations, reducing the need for manual review of every recording. These programs use algorithms to identify specific patterns in the sound waves, allowing you to quickly filter out irrelevant data and focus on the calls of interest. Furthermore, maintaining detailed metadata, such as the date, time, location, and equipment settings for each recording, is essential for accurate analysis and future reference.

After collecting and processing your data, the next step is to interpret the results and draw meaningful conclusions. This requires not only a strong understanding of acoustics but also a solid foundation in ecology and animal behavior. For example, an increase in the number of frog calls during a particular season could indicate a successful breeding period, while a decrease in bird songs could signal habitat loss or environmental degradation. By combining your acoustic data with other sources of information, such as visual observations, weather patterns, and land use changes, you can gain a more holistic understanding of the ecological processes occurring within the monitored area.

Remember that wildlife monitoring is not just about collecting data; it's about using that data to inform conservation efforts and protect vulnerable species. The insights gained from remote listening can be used to identify critical habitats, assess the impact of human activities on wildlife populations, and track the effectiveness of conservation interventions. By sharing your findings with other researchers, policymakers, and the public, you can contribute to a broader understanding of the importance of biodiversity and the need for sustainable resource management. This collective knowledge empowers us to make informed decisions that benefit both wildlife and the environment.

One aspect often overlooked is the importance of community engagement. Local communities often possess invaluable knowledge about the local ecosystem and can play a vital role in the success of your monitoring program. By involving them in data collection, analysis, and interpretation, you can foster a sense of ownership and responsibility for the conservation of their natural resources. This collaborative approach not only enhances the scientific rigor of your work but also promotes a more inclusive and equitable approach to conservation. Consider organizing

workshops, training sessions, or citizen science projects to empower community members to actively participate in wildlife monitoring efforts.
Think about the future of wildlife monitoring. The advancements in remote listening technology are constantly pushing the boundaries of what's possible. From miniaturized recording devices that can be deployed in the most remote locations to sophisticated artificial intelligence algorithms that can automatically identify and classify animal vocalizations, the possibilities are endless. As these technologies continue to evolve, they will undoubtedly play an increasingly important role in our efforts to understand and protect the world's biodiversity. Embrace these advancements and integrate them into your monitoring programs to gain even deeper insights into the complex lives of animals and the ecosystems they inhabit.
Ultimately, wildlife monitoring is a long-term commitment that requires dedication, perseverance, and a genuine passion for the natural world. It's not always easy; there will be challenges along the way, from equipment malfunctions to unexpected environmental events. However, the rewards of contributing to the conservation of biodiversity and the preservation of our planet's natural heritage are immeasurable. So, step out into the wild, listen carefully, and let the sounds of nature guide your journey. The silent cairn whispers its secrets to those who are willing to listen.

6.3 Medical Diagnostics

Auscultation, the practice of listening to internal body sounds, has long been a cornerstone of medical diagnosis. From the familiar whoosh of blood flow to the subtle crackles in the lungs, these sounds provide invaluable clues about a patient's condition. But traditional auscultation, reliant on a stethoscope and direct contact, can be limiting, especially in remote areas or situations where physical interaction is risky. Remote listening technologies, the same that are employed in surveillance and wildlife monitoring, offer a transformative approach to medical diagnostics. Imagine doctors in urban centers being able to monitor a patient in a rural hospital, or specialists able to review the data from their home,

vastly improving the accessibility and quality of healthcare.
The adaptation of remote listening technology to the medical field involves several key stages. First, high-quality microphones, sometimes embedded in wearable sensors, are used to capture the relevant body sounds. These microphones need to be sensitive enough to pick up faint sounds while minimizing background noise, a considerable challenge in busy hospital environments. This is a crucial point since the initial data dictates the quality of the resulting information. Then comes the signal processing stage, where sophisticated algorithms filter, amplify, and analyze the captured sounds. This involves removing extraneous noises like air conditioning hum or speech and enhancing the desired sounds like heart murmurs or lung sounds. Techniques like noise cancellation and spectral analysis play a crucial role here, allowing clinicians to focus on the essential information. Finally, the processed audio data is transmitted securely to a remote location, where it can be accessed and analyzed by medical professionals. The data must be protected and delivered in a fast and efficient manner to have practical relevance.
One of the most promising applications of remote listening technologies in medicine is in the field of cardiology. Heart sounds provide a wealth of information about the heart's structure and function. Remote auscultation can be used to detect heart murmurs, abnormal valve function, and other cardiac abnormalities. This is particularly useful for patients in rural areas who may not have access to specialized cardiac care. Imagine a scenario where a general practitioner in a remote clinic can use a digital stethoscope to record a patient's heart sounds and then transmit the data to a cardiologist in a major city for review. This can lead to earlier diagnosis and treatment of heart conditions, potentially saving lives. In addition to diagnosis, remote monitoring of heart sounds can also be used to track the progression of heart disease and assess the effectiveness of treatment. Patients with chronic heart conditions can be monitored remotely, allowing for early detection of any deterioration and prompt intervention. This can reduce the need for hospitalizations and improve the overall quality of life for patients with heart disease.
Pulmonary medicine is another area where remote listening technologies hold immense promise. Lung sounds can reveal a variety of respira-

tory conditions, such as pneumonia, asthma, and chronic obstructive pulmonary disease (COPD). Remote auscultation can be used to detect abnormal lung sounds, such as wheezes, crackles, and rhonchi, which can indicate the presence of these conditions. This is particularly valuable for patients with chronic respiratory diseases who need regular monitoring. For example, patients with COPD can use a wearable sensor to monitor their lung sounds at home. The data can then be transmitted to their physician, who can track their condition and adjust their treatment plan as needed. This can help prevent exacerbations of COPD and improve the patient's overall respiratory health. Moreover, remote listening can be particularly useful in infectious disease monitoring, helping to assess the respiratory effects of illnesses like influenza or emerging viruses. This capability becomes critical during pandemics, when in-person assessments may be limited and the need for remote monitoring is paramount.

Beyond cardiology and pulmonology, remote listening technologies are finding applications in other areas of medicine. For example, they can be used to monitor bowel sounds in patients with gastrointestinal disorders. Abnormal bowel sounds can indicate the presence of a bowel obstruction, inflammation, or other gastrointestinal problems. Remote monitoring of bowel sounds can help physicians diagnose and manage these conditions more effectively. In addition, remote listening can be used to monitor fetal heart sounds during pregnancy. This can provide reassurance to expectant mothers and allow for early detection of any potential problems. These technologies are also being explored for use in monitoring muscle sounds and joint sounds, potentially aiding in the diagnosis and management of musculoskeletal disorders. The potential applications are vast and continue to expand as the technology advances.

Despite the immense potential of remote listening technologies in medical diagnostics, there are several challenges that need to be addressed. One of the biggest challenges is ensuring the accuracy and reliability of the captured audio data. Background noise, variations in microphone placement, and individual differences in body size and shape can all affect the quality of the audio signal. Sophisticated signal processing algorithms are needed to overcome these challenges and extract the relevant information from the noise. Another challenge is ensuring the security and privacy of the transmitted audio data. Medical information is highly sensitive, and it is essential to protect it from unauthorized access. Robust encryption and security protocols are needed to ensure the confidentiality of patient data. Moreover, regulatory hurdles and ethical considerations need careful attention. Issues related to data ownership, consent, and liability need to be addressed before these technologies can be widely adopted in clinical practice.

However, addressing these challenges unlocks a new era of patient care. It can increase access to specialized medical expertise, especially in underserved communities. It empowers patients to actively participate in their healthcare management by monitoring their own body sounds and sharing the data with their physicians. Remote listening technologies also pave the way for innovative approaches to medical education and training. Medical students can use remote auscultation devices to listen to real-time body sounds from patients in different locations, enhancing their diagnostic skills and expanding their clinical experience. By overcoming the existing challenges and embracing the opportunities, remote listening technologies can revolutionize medical diagnostics and improve healthcare outcomes for patients worldwide. The future of medicine involves a fusion of traditional clinical expertise with cutting-edge acoustic technology, and this fusion holds tremendous promise for transforming the way we diagnose and treat diseases.

6.4 Assistive Listening Devices

Imagine being surrounded by a world teeming with sounds you can't quite grasp. This is the reality for millions experiencing hearing loss, a reality that assistive listening devices (ALDs) strive to bridge. These devices aren't just about making sounds louder; they're about enhancing clarity, reducing background noise, and tailoring the listening experience to individual needs and specific environments. Let's delve into the world of ALDs, exploring their diverse types and their profound impact on those who depend on them.

There exists a variety of ALDs, each designed to address specific listening challenges. Among the most prevalent are hearing aids, sophisticated miniaturized electronic devices that amplify sound. But hearing aids are only the start. Personal amplifiers provide similar amplification but are typically simpler and more affordable. Beyond amplification, frequency modulation (FM) systems transmit audio signals wirelessly from a speaker to a receiver worn by the listener, overcoming distance and background noise issues. Infrared systems function similarly, using infrared light to transmit the signal. Induction loop systems, also known as hearing loops, utilize a magnetic field to transmit audio from a sound source directly to a hearing aid equipped with a telecoil (t-coil). Finally, alerting devices use visual or tactile cues, such as flashing lights or vibrations, to signal important sounds like doorbells, alarms, or phone calls. Selecting the right ALD depends critically on the type and extent of hearing loss, as well as the specific listening situations where assistance is needed.

FM systems have a unique place in educational settings and other environments where distance and background noise are significant obstacles. Imagine a student with hearing loss trying to follow a teacher's lecture in a bustling classroom. The teacher wears a microphone connected to a transmitter, while the student wears a receiver connected to their hearing aid or headphones. The FM system wirelessly transmits the teacher's voice directly to the student, effectively bypassing the surrounding noise and distance, ensuring clear and focused listening. A key advantage of FM systems is their portability, allowing them to be used in various locations. Furthermore, certain FM systems offer multiple channels, enabling several speakers to be used simultaneously without causing interference. They are a valuable tool to improve communication in dynamic and noisy environments.

Another type of ALD, infrared systems, operates on a similar principle to FM systems but use infrared light instead of radio waves. These systems are commonly found in theaters, cinemas, and concert halls, providing clear audio directly to the listener. Unlike FM systems, infrared systems are less susceptible to interference from other electronic devices, which may make them better suited in environments with high radio frequency activity. However, infrared signals are easily blocked by objects, limiting their range and requiring a direct line of sight between the transmitter and the receiver. Despite this limitation, infrared systems offer a reliable and secure way to deliver audio in designated spaces.

Induction loop systems, also known as hearing loops, work by creating a magnetic field that can be picked up by a telecoil (t-coil) in a hearing aid. A wire loop is installed around a room or area, such as a ticket counter, and connected to an amplifier and a sound source. When the sound source is activated, it generates a magnetic field within the loop. Individuals with hearing aids equipped with a t-coil can switch their hearing aid to the t-coil setting, which allows the hearing aid to receive the audio signal directly from the magnetic field, bypassing background noise. Hearing loops are increasingly common in public spaces such as churches, theaters, and government buildings, offering a discreet and effective way for people with hearing loss to access audio information. However, to benefit from a hearing loop, individuals must have a hearing aid with a t-coil, and the loop must be properly installed and maintained to ensure optimal performance.

Alerting devices represent a different class of ALDs, focusing on providing non-auditory alerts for important sounds. These devices use visual or tactile cues to notify individuals of events such as a ringing doorbell, a smoke alarm, or a crying baby. Visual alerting devices often employ flashing lights, while tactile devices use vibrations. Alerting devices are particularly useful for individuals with severe to profound hearing loss who may not be able to hear traditional auditory alerts. These devices are incredibly useful for daily life. They can significantly enhance safety and independence by ensuring that individuals with hearing loss are aware of important events and potential hazards in their environment.

When advising someone on choosing an ALD, several crucial factors must be considered. An audiologist's comprehensive hearing evaluation serves as the starting point. This evaluation will determine the type and degree of hearing loss, as well as the individual's specific listening needs. The audiologist can then recommend specific ALDs that are appropriate for the individual's hearing loss and lifestyle. The environment in which the ALD will be used is another critical consideration. For example, an FM system may be ideal for a student in a noisy classroom, while an induction loop may be more suitable for someone who frequently attends theater performances. Cost is also a factor, as ALDs can range in price from a few dollars for a simple personal amplifier to several thousand dollars for a sophisticated hearing aid. Furthermore, user-friendliness is essential, especially for older adults or those with limited dexterity. The ALD should be easy to use, adjust, and maintain.

The impact of ALDs extends far beyond simply making sounds louder. For individuals with hearing loss, ALDs can dramatically improve communication, participation, and quality of life. They can enable individuals to participate more fully in conversations, understand lectures and presentations, enjoy music and entertainment, and feel more connected to the world around them. For children with hearing loss, ALDs can play a crucial role in language development and academic success. By providing access to clear and consistent auditory information, ALDs can help children learn to speak, read, and write. The ability to communicate effectively is a fundamental human need, and ALDs can help individuals with hearing loss overcome the barriers that hearing loss can create. They can improve social interactions, reduce feelings of isolation, and enhance self-esteem.

As technology advances, we can expect to see even more innovative and effective ALDs emerge. Improved hearing aids, more sophisticated FM and infrared systems, and more versatile alerting devices are already on the horizon. The future of ALDs is bright, with the potential to transform the lives of millions of people with hearing loss, granting them the opportunity to engage more fully and meaningfully with the world around them. By continually developing and refining these technologies, we can help ensure that everyone has the opportunity to experience the richness and beauty of sound.

6.5 Security Protocols

Securing remote listening systems is paramount, particularly given their potential for misuse and the sensitive nature of the audio data they handle. Therefore, robust security protocols are essential to protect both the integrity of the system and the privacy of individuals who may be inadvertently recorded.

The foundation of any security protocol lies in strong authentication mechanisms. This means verifying the identity of anyone attempting to access the system, whether it's an administrator managing the system or a user retrieving audio data. Simple passwords are no longer sufficient; multi-factor authentication (MFA) should be mandatory. MFA requires users to provide two or more verification factors, such as something they know (password), something they have (a security token or mobile app), or something they are (biometric data like a fingerprint). This drastically reduces the risk of unauthorized access, even if one factor is compromised. Beyond MFA, consider implementing role-based access control (RBAC). RBAC restricts system access based on a user's role within the organization. For example, a technician might have access to system settings but not to archived audio data, while an analyst might have the opposite permissions. Regular audits of user accounts and permissions are crucial to ensure that access levels remain appropriate as roles evolve. Implementing a "least privilege" principle helps. This principle states that a user should only have the minimum level of access needed to perform their job. This minimizes the potential damage if an account is compromised.

Encryption is another cornerstone of audio data security. Audio data should be encrypted both in transit and at rest. In transit, secure protocols like Transport Layer Security (TLS) should be used to encrypt communication between the microphone, the server, and any client devices. This prevents eavesdropping during data transmission. Encryption at rest protects the audio data when it's stored on servers or other storage devices. AES-256 is a widely respected encryption algorithm that provides a high level of security. Implement key management procedures to securely store and manage the encryption keys. Poor key management can render even the strongest encryption useless. Ensure compliance with relevant encryption standards and regulations, such as HIPAA for medical data or GDPR for personal data. This helps to minimize liability and maintain public trust.

Network security is a critical component of the overall security posture. The network on which the remote listening system operates must be properly secured to prevent unauthorized access and data breaches. Deploy firewalls to control network traffic and prevent unauthorized connections.

Intrusion detection and prevention systems (IDS/IPS) can detect and block malicious activity on the network. Segment the network to isolate the remote listening system from other parts of the network. This limits the impact of a security breach if one occurs. Regularly scan the network for vulnerabilities and patch any security holes promptly.

Physical security is sometimes overlooked, but it's an important aspect of protecting remote listening systems. Protect the physical devices, such as microphones and servers, from theft or tampering. Secure the locations where these devices are installed with physical access controls, such as locks, surveillance cameras, and alarm systems. Implement procedures to control physical access to the server room or data center where the audio data is stored. Require background checks for employees who have access to sensitive data.

Regularly test and audit the security protocols to ensure that they are effective. Conduct penetration tests to identify vulnerabilities in the system. Review security logs regularly to look for suspicious activity. Perform security audits to assess the overall security posture of the system.

Data retention policies are important for managing the lifecycle of audio data and reducing the risk of data breaches. Establish clear data retention policies that define how long audio data will be stored. Securely dispose of audio data when it's no longer needed. Implement data masking or anonymization techniques to protect sensitive information in audio data. Train employees on data security best practices and the importance of protecting sensitive information. Consider using watermarking techniques to embed identifying information into the audio data. This can help to track the source of a leak if the data is compromised.

Incident response planning is crucial for handling security breaches and other security incidents. Develop an incident response plan that outlines the steps to be taken in the event of a security breach. Assign roles and responsibilities to individuals who will be involved in the incident response process. Practice the incident response plan regularly through simulations and exercises.

Compliance with regulations and standards is essential for ensuring that remote listening systems are secure and compliant with applicable laws and regulations. Research and understand the relevant regulations and standards, such as GDPR, HIPAA, and PCI DSS. Implement security controls to comply with these regulations and standards. Regularly audit the system to ensure that it remains in compliance.

Vendor security is an important consideration when using third-party services or software in a remote listening system. Assess the security practices of vendors before engaging with them. Include security requirements in contracts with vendors. Regularly monitor vendor security practices to ensure that they remain adequate.

Continuous monitoring and improvement is key to maintaining the security of remote listening systems over time. Continuously monitor the system for security threats and vulnerabilities. Regularly update security protocols to address new threats. Stay informed about the latest security best practices and technologies.

By implementing these comprehensive security protocols, it is possible to significantly reduce the risk of unauthorized access, data breaches, and other security incidents, ensuring the responsible and ethical use of remote listening technology. This is especially important given the sensitive nature of audio data and the potential for privacy violations. Furthermore, the establishment of robust security protocols helps to build trust in the system, encouraging its adoption and acceptance. Remember that security is not a one-time task but an ongoing process that requires constant vigilance and adaptation.

7 Chapter 7: Ethical Considerations

As remote listening technologies advance, the ethical landscape surrounding their use becomes increasingly complex, demanding careful consideration and proactive measures. The ability to capture, transmit, and analyze audio from a distance presents unparalleled opportunities, but also poses significant risks to privacy, data security, and individual autonomy. Navigating this terrain requires a commitment to ethical principles, a thorough understanding of potential harms, and the implementation of robust safeguards.

Privacy, perhaps the most prominent concern, is directly challenged by remote listening capabilities. The very act of recording audio, even in public spaces, can be perceived as intrusive and violate an individual's expectation of privacy. Consider the placement of microphones in seemingly innocuous locations, such as parks or public transportation, without clear signage or public consultation. While proponents might argue that such surveillance is necessary for security purposes, the potential for misuse and the chilling effect on free expression cannot be ignored. Imagine a scenario where individuals are hesitant to engage in political discussions or express dissenting opinions for fear of being overheard and potentially targeted. This erosion of privacy can have a profound impact on democratic discourse and social cohesion.

Data security is another critical aspect that demands attention. The storage and transmission of audio data, particularly when it contains sensitive information, are vulnerable to breaches and unauthorized access.

Think about medical diagnostics, where remote auscultation devices are used to monitor patients' heart and lung sounds. If this data is not adequately protected, it could be intercepted by malicious actors and used for identity theft, blackmail, or other nefarious purposes. Similarly, in wildlife monitoring applications, audio recordings might reveal the presence of endangered species or sensitive habitats, information that could be exploited by poachers or developers. Implementing robust encryption protocols, access controls, and data retention policies is essential to mitigate these risks and ensure that audio data is handled responsibly. The technical expertise to build the encryption system is not always available, and it requires specific training.

Responsible innovation is not just a slogan, but a guiding principle that should inform the entire lifecycle of remote listening technologies, from design and development to deployment and evaluation. This requires a multidisciplinary approach, involving not only engineers and scientists, but also ethicists, legal experts, and members of the communities that may be affected. A key component of responsible innovation is transparency. Organizations deploying remote listening technologies should be upfront about their purpose, scope, and potential impact. They should also provide clear and accessible information about how data is collected, stored, and used, and offer individuals the opportunity to opt out or access their own data. Public awareness campaigns and educational initiatives can play a vital role in fostering informed consent and building trust.

Furthermore, responsible innovation necessitates a commitment to minimizing harm. This includes carefully considering the potential for bias in algorithms used to analyze audio data. For example, speech recognition systems have been shown to be less accurate for certain accents or dialects, which could lead to unfair or discriminatory outcomes. Similarly, emotion recognition technologies, which claim to detect emotions from vocal cues, are often based on flawed assumptions and lack scientific validity. Deploying such technologies without proper validation and oversight could have serious consequences, particularly in high-stakes contexts such as law enforcement or employment.

Accountability is another cornerstone of ethical remote listening. There must be clear lines of responsibility for ensuring that these technologies are used in a way that respects individual rights and societal values. This may involve establishing independent oversight bodies or ethics review boards to monitor the use of remote listening technologies and investigate complaints of abuse. It also requires holding individuals and organizations accountable for violations of privacy or data security. Strong legal frameworks and regulatory mechanisms are essential to deter unethical behavior and ensure that there are consequences for those who misuse these powerful tools. We need to be aware and use the technologies carefully.

Considering the potential for misuse, such as the use of microphones in toys or household appliances without the user's knowledge or consent, is crucial. These hidden microphones could be used to eavesdrop on conversations, collect personal information, or even manipulate the user's behavior. To prevent such abuses, manufacturers should be required to disclose the presence of microphones in their products and obtain explicit consent from users before activating them. Furthermore, it is essential to develop technical solutions to detect and disable unauthorized microphones. Thinking about the consequences of the use of a new technology makes us aware of its dangers.

In conclusion, the ethical considerations surrounding remote listening technologies are multifaceted and demand a proactive and responsible approach. By prioritizing privacy, data security, and responsible innovation, we can harness the power of these technologies for good while mitigating their potential harms. This requires a collective effort from engineers, policymakers, ethicists, and the public to ensure that remote listening technologies are used in a way that aligns with our values and promotes a just and equitable society.

7.1 Privacy Concerns

The insidious creep of technology into our lives often occurs under the guise of convenience or security, yet with each advancement, the line between helpful tool and intrusive surveillance blurs. Remote listening technologies, with their capacity to capture and transmit sound from distant locations, represent a particularly potent double-edged sword. The potential for misuse, whether intentional or accidental, is significant and demands careful consideration.

One of the primary privacy concerns stems from the very nature of audio data. Unlike other forms of data, such as location information or browsing history, audio recordings capture not just the content of conversations, but also the nuances of tone, emotion, and identity. They can reveal sensitive personal details, from medical conditions hinted at in coughs and sneezes to private discussions about finances, relationships, or political beliefs. This richness of information makes audio data incredibly valuable – and incredibly vulnerable. Consider the implications of a security camera with audio recording capabilities installed in a public park. While intended to deter crime, it also captures the private conversations of families, friends, and couples, creating a chilling effect on free expression and casual interaction. The question then becomes: what oversight is in place to ensure appropriate use and prevent abuse? Who determines what constitutes a legitimate need for audio surveillance, and what mechanisms are in place to hold those accountable who overstep these boundaries?

Further complicating matters is the increasing sophistication of audio analysis techniques. What was once considered background noise can now be parsed, analyzed, and even reconstructed into intelligible conversations. Sophisticated algorithms can identify speakers, transcribe speech, and even infer emotional states from voice patterns. This means that even seemingly innocuous audio recordings can be mined for a wealth of personal information, often without the knowledge or consent of the individuals being recorded. Think about the smart home devices marketed as helpful assistants. They constantly listen for voice commands, but what happens to the audio data they collect? Is it stored securely? Is it used for targeted advertising? Is it shared with third parties? These questions are not easily answered, and the lack of transparency surrounding data collection and usage practices is a major cause for concern. I believe we need to have very clear about how these devices operate in our private space.

The potential for mass surveillance is another significant threat. Imagine a city equipped with an extensive network of microphones, ostensibly for monitoring traffic noise or detecting gunshots. While such a system might have legitimate public safety benefits, it also creates the potential for indiscriminate and continuous audio surveillance of the entire population. This kind of pervasive monitoring can have a chilling effect on freedom of speech and assembly, as individuals may be less likely to express dissenting opinions or participate in protests if they know they are being constantly monitored. Even if the data is not actively monitored, the mere existence of such a system can create a climate of fear and self-censorship. The risk extends beyond governments; corporations could deploy similar technologies to monitor employee behavior, track customer interactions, or even eavesdrop on competitors. The scale and scope of this potential surveillance are staggering, and the implications for privacy and civil liberties are profound.

Addressing these privacy concerns requires a multi-faceted approach. Strong legal frameworks are essential to regulate the collection, storage, and use of audio data. These laws should clearly define the circumstances under which audio surveillance is permissible, limit the types of data that can be collected, and establish strict safeguards to prevent abuse. Transparency is also crucial. Companies should be required to disclose their audio data collection practices in clear and accessible language, and individuals should have the right to access, correct, and delete their audio data. Furthermore, individuals need to be empowered to protect their own privacy. This includes educating them about the risks of audio surveillance and providing them with tools and techniques to mitigate those risks, such as using signal jammers, encryption, or simply being more mindful of what they say in public spaces. We should be educated about the impact of our behaviors.

Ultimately, the ethical use of remote listening technologies depends on a fundamental shift in perspective. We need to move away from the notion that technology should be deployed simply because it is possible, and instead adopt a more cautious and considered approach that prioritizes privacy and human rights. This requires a collaborative effort involving technologists, policymakers, and the public to develop ethical guidelines, promote responsible innovation, and ensure that these powerful technologies are used for the benefit of society, rather than to its detriment. I feel we need to be careful to not overstep boundaries and respect individual privacy rights. It's a fine balance but definitely worth striving for.

Finally, we must also consider the impact of remote listening technologies on vulnerable populations. Children, the elderly, and individuals with disabilities are particularly susceptible to exploitation and abuse through these technologies. For example, caregivers might use remote listening devices to monitor the behavior of their charges, but this could easily cross the line into intrusive surveillance and violate their privacy rights. Similarly, law enforcement agencies might target marginalized communities with audio surveillance, exacerbating existing inequalities and discrimination. Protecting these vulnerable groups requires targeted interventions, such as stricter regulations on the use of remote listening technologies in care settings and enhanced oversight of law enforcement practices.

7.2 Data Security Issues

Audio data, unlike other forms of information, possesses an inherent sensitivity due to its potential to capture not only spoken words but also contextual sounds that reveal intimate details about an individual's activities, environment, and even emotional state. Consider the seemingly innocuous recording of a bird song; if that recording also captures a snippet of a private conversation, the privacy implications immediately escalate. Thus, ensuring the robust security of audio data is not merely a technical challenge but a fundamental ethical imperative. Securing audio data presents a multifaceted challenge, demanding a layered approach that spans from the point of capture to storage, transmission, and ultimate disposal. Ignoring any part of this chain can weaken the overall protection.

The point of capture represents the first and arguably most vulnerable point in the audio data lifecycle. Microphones, the primary transducers in the remote listening chain, are often deployed in environments with limited physical security, making them susceptible to tampering and unauthorized access. A rogue actor gaining control of a microphone can not only intercept audio but also potentially inject malicious signals or disable the device altogether. Secure microphone designs should incorporate tamper-evident features and robust authentication mechanisms to prevent unauthorized access. Furthermore, the transmission of audio from the microphone to a central processing unit should be encrypted to prevent eavesdropping. Data encryption is a good start.

Data transmission is a critical juncture where audio data is most susceptible to interception. Wireless transmission, while convenient, introduces vulnerabilities that must be carefully addressed. Standard Wi-Fi networks, unless properly configured, can be easily compromised, allowing attackers to sniff network traffic and capture audio data. Similarly, Bluetooth connections are susceptible to eavesdropping and man-in-the-middle attacks. Securing wireless audio transmission requires the implementation of strong encryption protocols such as Advanced Encryption Standard (AES) with sufficiently long keys. In addition, secure authentication mechanisms should be employed to ensure that only authorized devices can access the audio stream. Virtual Private Networks (VPNs) can add another layer of protection, creating an encrypted tunnel for audio transmission over public networks. Regular security audits and penetration testing can help identify vulnerabilities in wireless audio transmission systems. Don't be afraid to test the technology often.

Storage of audio data poses a persistent security risk, as data at rest is a prime target for cyberattacks. Centralized data storage systems, while offering scalability and accessibility, also represent a single point of failure. Breaches of these systems can result in the mass compromise of sensitive audio data. To mitigate this risk, audio data should be encrypted both in transit and at rest. Access controls should be strictly enforced, limiting access to only authorized personnel. Regular backups should be maintained, but these backups must also be securely stored and encrypted. Distributed storage systems, where data is spread across multiple locations, can enhance security by reducing the impact of a single point of failure. Secure deletion of audio data is equally important, ensuring that sensitive information cannot be recovered after it is no longer needed. Overwriting data with random characters multiple times can effectively prevent data recovery. Consider encrypting and shredding the data if possible.

Beyond technical safeguards, robust data governance policies are essential for ensuring the security of audio data. These policies should clearly define who has access to audio data, how it can be used, and how long it should be retained. Regular training should be provided to personnel who handle audio data to ensure that they understand and adhere to these policies. Data breach response plans should be in place to outline the steps to be taken in the event of a security incident. Compliance with relevant data privacy regulations, such as the General Data Protection Regulation (GDPR), is crucial for maintaining trust and avoiding legal penalties. Transparency with individuals about how their audio data is being collected, used, and protected is also essential for fostering ethical and responsible use of remote listening technology. Always make sure you are transparent about the audio.

Moreover, the choice of audio codecs and processing algorithms can inadvertently introduce security vulnerabilities. Certain codecs may have known weaknesses that can be exploited by attackers to inject malicious code or manipulate audio data. Similarly, flawed processing algorithms can lead to unintended consequences, such as amplifying background noise or distorting speech patterns, potentially compromising the integrity of the audio information. Selecting codecs and algorithms that have been thoroughly vetted for security vulnerabilities is crucial. Regular updates and patches should be applied to address any newly discovered vulnerabilities. Employing cryptographic techniques, such as digital watermarking, can help detect tampering and ensure the authenticity of audio data. Be sure to choose codecs that prioritize security as well as performance.

Consider the supply chain risks associated with remote listening technology. Microphones, audio processing units, and network devices are often sourced from multiple vendors, each with their own security practices. A vulnerability in any component of the supply chain can compromise the security of the entire system. Thoroughly vetting vendors and ensuring that they adhere to industry best practices for security is essential. Requiring vendors to provide security certifications and conducting regular audits can help identify and mitigate supply chain risks. Implementing strong authentication and authorization mechanisms can prevent unauthorized devices from accessing the audio network. Consider where the

equipment is coming from.
Emerging technologies, such as edge computing and federated learning, offer new opportunities for enhancing the security of audio data. Edge computing allows audio processing to be performed locally on the device, reducing the need to transmit raw audio data over the network. This can significantly reduce the risk of interception and improve data privacy. Federated learning enables machine learning models to be trained on distributed audio data without requiring the data to be centralized. This can enable new insights to be extracted from audio data while preserving data privacy. Exploring and adopting these emerging technologies can help organizations enhance the security and privacy of their remote listening systems. Keep up to date with the technology.
In conclusion, securing audio data requires a comprehensive and proactive approach that addresses vulnerabilities at every stage of the data lifecycle. By implementing robust technical safeguards, establishing sound data governance policies, and staying abreast of emerging threats and technologies, organizations can protect the privacy of individuals and ensure the responsible use of remote listening technology. Never stop improving the security.

7.3 Responsible Innovation

The power of remote listening technology comes with a profound responsibility. As we develop and deploy systems capable of capturing and transmitting audio from increasingly distant locations, the potential for misuse grows exponentially. Responsible innovation, therefore, demands a proactive and multifaceted approach, one that considers not only the technical capabilities of these systems but also their social, ethical, and legal implications. It is about more than simply adhering to existing laws; it requires anticipating future challenges and shaping the technology's development to align with our values.
One crucial aspect of responsible innovation is prioritizing privacy from the outset. This means designing systems with privacy-enhancing technologies (PETs) built-in, rather than treating privacy as an afterthought. This may involve using techniques such as differential privacy, which

adds statistical noise to audio data to obscure individual identities while still allowing for meaningful analysis. Another approach is federated learning, where algorithms are trained on decentralized audio data without the need to centralize the data itself, thus reducing the risk of data breaches and privacy violations. End-to-end encryption is also critical for protecting audio data during transmission and storage, ensuring that only authorized parties can access the information. We can consider the design of microphone arrays, which should include features to limit the directionality and range of audio capture when appropriate, avoiding the unintentional recording of conversations outside the intended area. Transparency is key; users should be clearly informed when they are being recorded, and they should have the right to access, correct, and delete their audio data, in accordance with data protection regulations.

Beyond privacy, data security is paramount. Remote listening systems often generate vast amounts of audio data, which can be vulnerable to cyberattacks and unauthorized access. Robust security protocols are essential to protect this data from falling into the wrong hands. This includes implementing strong authentication mechanisms, such as multi-factor authentication, to prevent unauthorized access to the system. Regular security audits and penetration testing can help identify and address vulnerabilities before they can be exploited. Data minimization is another important principle; only the minimum amount of audio data necessary for the intended purpose should be collected and stored, reducing the potential impact of a data breach. We need to consider secure storage solutions, such as encrypted databases and cloud storage services with robust security certifications. Furthermore, incident response plans should be in place to quickly detect and respond to any security breaches, minimizing the damage and notifying affected individuals and organizations.

The application of remote listening technology raises unique ethical challenges. Consider, for example, the use of these systems for surveillance purposes. While surveillance can be a valuable tool for law enforcement and security agencies, it can also be used to suppress dissent, discriminate against marginalized groups, and chill freedom of expression. To mitigate these risks, clear legal frameworks are needed to govern the use of remote listening technology for surveillance. These frameworks should specify the circumstances under which surveillance is permitted, the safeguards that must be in place to protect individual rights, and the oversight mechanisms to ensure accountability. We can encourage the development of ethical guidelines for the use of remote listening technology in different contexts, involving stakeholders from diverse backgrounds, including ethicists, legal scholars, technologists, and community representatives. Education and training are crucial to ensure that those who develop and deploy these systems are aware of the ethical considerations and are equipped to make responsible decisions.

Responsible innovation also requires ongoing dialogue and engagement with the public. It is essential to foster a broader understanding of the capabilities and limitations of remote listening technology, as well as the potential risks and benefits. Public forums, workshops, and online resources can help to inform the public and solicit feedback on the ethical and social implications of these technologies. We can actively engage with community organizations and advocacy groups to address concerns and build trust. This proactive engagement can help to shape the development of remote listening technology in a way that is aligned with public values and expectations. Moreover, it's crucial to continuously monitor and evaluate the impact of remote listening technology on society. This involves tracking both the intended and unintended consequences of these systems, identifying potential harms, and adapting policies and practices accordingly.

Ultimately, responsible innovation in the realm of remote listening technology is not a destination but a journey. It requires a continuous commitment to ethical principles, a willingness to learn and adapt, and a dedication to ensuring that these powerful tools are used for the benefit of society as a whole. The quest goes beyond mere compliance; it embodies a dedication to crafting technology that mirrors our highest ideals. We must strive for an approach where technology enhances, rather than diminishes, the fundamental rights and freedoms we cherish. Only through such sustained effort can we navigate the complex ethical terrain and unlock the full potential of remote listening technology in a way that is both innovative and responsible.

This includes developing technologies that promote transparency and accountability in the use of these systems, such as audit trails and data provenance mechanisms. Investing in research on the social and ethical implications of remote listening technology can help to inform policy decisions and guide the development of responsible innovation frameworks. Encouraging collaboration between researchers, industry, government, and civil society can foster a more holistic and informed approach to responsible innovation. The future of remote listening technology depends on our ability to address these ethical challenges proactively and thoughtfully, ensuring that it serves as a force for good in the world.

www.ingramcontent.com/pod-product-compliance
Lightning Source LLC
LaVergne TN
LVHW022017200926
848481LV00007B/2435
* 9 7 9 8 3 4 9 5 1 9 1 5 4 *